Infinity Paradoxes

Infinity Paradoxes

A Down-to-Earth Guide to Mind-Blowing Math

Avery Pascal

Avery Pascal

Copyright 2024 Avery Pascal. All Rights reserved. No part of this publication may be reproduced without the consent of the author.

Infinity Paradoxes

"Two things are infinite: the universe and human stupidity; and I'm not sure about the universe."

— **Albert Einstein**

Avery Pascal

Table of Contents

Chapter 1: Hilbert's Hotel – Infinite Rooms, Infinite Guests

Chapter 2: Cantor's Diagonal Argument – When Infinity Comes in Sizes

Chapter 3: Zeno's Paradoxes – The Race That Never Ends

Chapter 4: Thompson's Lamp – Flickering Forever in Finite Time

Chapter 5: Gabriel's Horn – A Finite Volume with Infinite Surface

Chapter 6: The Ross-Littlewood Paradox – The Vase That Never Fills

Chapter 7: The Banach-Tarski Paradox – Doubling a Ball Without Breaking Rules

Chapter 8: The Infinite Monkey Theorem – Typing Shakespeare by Chance

Chapter 9: The Dartboard Paradox – Zero Chances but Certain Outcomes

Chapter 10: The St. Petersburg Paradox – Betting on Infinity

Introduction

Have you ever stared at the night sky and felt a sense of awe at its vastness? That's how I felt when I first encountered the concept of infinity. As a kid, I'd lie in my backyard, gazing at the stars, trying to wrap my head around the idea of something that goes on forever. It was mind-boggling, exciting, and a little scary all at once.

That childhood wonder never left me. It's what led me to write this book, "Infinity Paradoxes: A Down-to-Earth Guide to Mind-Blowing Math." But don't worry – you don't need to be a math whiz to join me on this adventure. This book is for anyone who's ever looked up at the stars and wondered, "What if?"

Infinity is like an endless ocean of possibilities. Every time we think we've reached its shore, we find there's more to explore. It's a concept that challenges our understanding of reality and pushes the boundaries of what we think is possible.

In the chapters ahead, we'll dive into some of the most fascinating paradoxes that arise when we try to grasp infinity. We'll meet a hotel with infinite rooms that can always accommodate one more guest, explore a set that contains all sets (including itself!), and even question whether infinity plus one is really bigger than infinity.

These aren't just abstract ideas – they have real-world implications that might surprise you. So, are you ready to challenge your assumptions and see the world in a new light? Let's set sail on this infinite ocean together. I promise it'll be a trip you won't forget!

Avery Pascal

Chapter 1: Hilbert's Hotel – Infinite Rooms, Infinite Guests

Infinite Accommodation

Picture a hotel that seems to stretch endlessly into the distance. It's a magnificent place with gleaming brass fixtures, soft carpets, and a lobby buzzing with people at all hours. This isn't your average hotel; it has an infinite number of rooms, each one labeled in order: Room 1, Room 2, Room 3, and so on, forever. Here's the twist: every single room is already filled with an infinite number of guests. Welcome to Hilbert's Hotel, where the idea of accommodation takes on a whole new meaning.

At first, the thought of a hotel packed to the brim might conjure images of tight spaces and long waits at the front desk. Think of a crowded hotel on a busy summer weekend, with families checking in, couples seeking a romantic escape, and business travelers gathered in the lobby, desperately hoping for that last available room. Now, contrast that scene with Hilbert's Hotel, a place where roominess is less of a problem and more of a fascinating concept. In this hotel, infinite rooms can hold infinite guests, and it all happens in a way that feels almost magical.

So, how does this endless accommodation actually work? Imagine a new guest arriving, excited to check in, only to find out that every room is occupied. You might think the staff would apologize and send them on their way, but things are different at Hilbert's Hotel. The clever staff has a plan: they ask each current guest to move to the next room over. The guest in Room 1 moves to Room 2, the guest in Room 2 moves to Room 3, and this pattern continues on and on. Miraculously, Room 1 opens up, and the new guest is happily checked in, all while the existing guests remain blissfully unaware of the shuffle.

This clever move beautifully illustrates the wonder of infinity—where you can always find room for more, even when it seems impossible. Thinking of it another way, it's like rearranging furniture in a small apartment to fit an unexpected guest. You might feel like you've run out of space, but with a little creativity and willingness to adapt, there's usually a way to squeeze in that extra chair or make a cozy nook for a sleeping bag.

We can find real-life examples of this concept everywhere. Think about a concert or a sporting event. Picture a scenario where the venue is completely full, yet the organizers still find a way to squeeze in one more person. They might offer standing-room-only tickets or, in a funny twist, set up a viewing area on a nearby rooftop. The key takeaway is clear: with a little imagination, there's always a way to make room.

Now, let's take it a step further. What happens when two new guests arrive at Hilbert's Hotel at the same time? Once again, the staff is ready with a solution. They can ask current guests to move into even-numbered rooms only, freeing up all the odd-numbered rooms for the newcomers. So, Room 1 becomes vacant, Room 3 becomes vacant, and so on. With the new guests comfortably settled into Rooms 1 and 3, it becomes clear that Hilbert's Hotel isn't just infinite; it's wonderfully accommodating.

As we think about these ideas, it's easy to fall back on our everyday experiences, which are often limited. We tend to think of finite numbers; a small restaurant can only have so many tables, and a cramped apartment can only fit so many people before someone starts worrying about safety. Yet, Hilbert's Hotel reminds us that infinity operates on a whole different level, where the usual rules of space and limits just don't apply.

Let's dive into the math behind this infinite hotel. In mathematics, there are different sizes of infinity, and Hilbert's Hotel introduces us to countable infinity, which means we can list the rooms (1, 2, 3, ...), just like we can list natural numbers. This strange property of infinity is foundational, and it challenges how we typically understand the world around us.

To make this point clearer, imagine if Hilbert's Hotel became a must-visit destination for mathematicians and curious explorers alike. Lines would form outside the grand entrance, with each guest excited to experience the wonder of infinite accommodation. Picture a group of mathematicians, clipboards and pens in hand, passionately discussing the implications of infinite sets while waiting to check in. Some might debate whether the hotel could genuinely house an infinite number of infinite guests. They would question how, if every room is full, more guests could possibly fit in. This lively debate highlights the paradox that Hilbert's Hotel presents—a place that showcases just how tricky and intriguing the idea of infinity can be.

Imagine the stories that would come out of guests who managed to stay at Hilbert's Hotel. There would be tales of late-night conversations in the lobby that stretch across time, where guests share their views on existence, infinity, and the universe. Maybe the person in Room 1 thinks of infinity as just a number that's too big to comprehend, while the occupant of Room 2 argues passionately that infinity is more about a state of being, a unique way of looking at everything that allows for endless possibilities.

The hotel would become a hub of philosophical discussions, unraveling the threads of logic and reality. Guests would dive into theories about infinity, ranging from the infinitely tiny (like a single atom) to the infinitely vast (the entire universe), making Hilbert's Hotel not just a stop for travelers, but also a sanctuary for thinkers.

While pondering the idea of infinite accommodation, we might chuckle at the notion of a guest who checks in and promptly gets lost in the maze of rooms. Picture them wandering from Room 1 to Room 2 to Room 3, only to discover they've forgotten which room was theirs. In this infinite space, the experience of roominess becomes fluid, an ever-shifting experience that defies the limits of our familiar understanding.

As we start to unravel the quirks of infinite accommodation, it becomes clear that this isn't just a whimsical story about a fantastical hotel. It's an open invitation to think deeply about existence, space, and the fabric of reality. It encourages us to push the boundaries of our understanding, to embrace the idea that life holds more than what can be measured or counted.

Ultimately, Hilbert's Hotel transforms into more than just an intellectual curiosity; it becomes a symbol of the possibilities that arise when we expand our minds beyond the usual constraints. Just as the hotel can always make room for new guests, we too can open our minds to the wealth of infinite ideas surrounding us—fostering creativity, challenging established norms, and broadening our view of the world. The notion of infinite accommodation is not just about the hotel; it's about exploring new realms of thought and discovering the limitless potential that lives within each of us.

Countable Infinity

When it comes to mathematics, the idea of infinity can feel overwhelming and a bit mysterious. But here's the good news: not all infinities are the same! Understanding the differences between them can really change how we think about numbers, sizes, and even our existence. So, let's unpack this fascinating concept called countable infinity.

First, let's look at the natural numbers: 1, 2, 3, 4, and so on. This set is infinite, but it has a special feature that makes it unique: you can actually count them! If you had an endless supply of numbers at your fingertips, you could keep ticking them off one by one.

There's a kind of order to it, like a line of numbers patiently waiting for you to recognize them. Mathematicians call this a "countably infinite" set, meaning it can be paired one-to-one with the natural numbers. This helps us organize infinity in a neat way.

Now, let's take a moment to appreciate how cool this is. Picture yourself matching each natural number with something tangible—like shoes, books, or guests in a never-ending hotel. The first shoe gets number 1, the second one gets number 2, and so forth. By doing this, we can visualize and understand our infinite collection. It's like hosting a party that never ends—everyone has a place to be, and there's always room for more guests. Countable infinity makes this kind of organization possible. But what about when we step outside the world of natural numbers?

This is where uncountable infinity comes into play, and it gets a bit mind-blowing. When we talk about uncountable sets, we refer to groups of numbers that can't be matched with natural numbers; they are "larger" than any countable infinity. A perfect example of this is the set of real numbers, which includes all fractions, decimals, and irrational numbers. Unlike natural numbers, which we can list (1, 2, 3...), real numbers stretch infinitely in a more tangled and dense way.

Let's make this clearer with a classic thought experiment called Cantor's diagonal argument. Imagine you're trying to list all the real numbers between 0 and 1. You might attempt to write them down like this:

0.1
0.12
0.123
0.1234
...

At first, it looks like you've got a complete list. But Cantor, the brilliant mind behind this idea, showed that no matter how thorough your list seems, you can always create a new real number that isn't on it. Just take the first digit from the first number, the second digit from the second number, and so on, then change each digit. This new number will be different from any number on your list by at least one digit. So, you can never fully capture all the real numbers between 0 and 1, proving it's uncountable.

And now, here's where it gets really exciting! Cantor's findings imply that there are not just different kinds of infinity; there's a whole hierarchy among them! Countable infinity is just a smaller part of the larger uncountable infinity. It's a bit like playing a video game with various levels; you start with easier ones that are fun and manageable, but as you level up, you encounter challenges that are infinitely more complex and captivating.

So how do we make sense of these two types of infinity? It's easy to think of infinity as a single vast thing, but mathematics shows us that the infinite world is full of variety. This realization not only deepens our understanding of numbers but also encourages us to think about how we see the world around us.

Consider the difference between finite and infinite. In our daily lives, we mostly deal with finite sets—like our bank accounts, the number of friends we have, or even our frustrations with traffic. These quantities are limited and can be easily counted. But when we step into the realm of countable and uncountable infinities, we're invited to break free from these limitations. The natural numbers, while infinite, feel like a cozy room in Hilbert's Hotel—there's a structure, a clear way to understand it, and a sense of how to fit more in.

In contrast, the uncountable infinity of real numbers resembles a vast, wild garden filled with countless flowers that you simply can't count or categorize. It's both dizzying and thrilling, showcasing the endless possibilities that lie beyond our usual numerical systems.

As we explore these concepts, let's think about how they challenge our intuitions. Picture this: you have an infinite number of invitations to a concert. If you send one invitation to each natural number, it seems like the venue is full. But what if someone asks for a plus-one? You can easily invite someone new by shifting the guests to different seats, just like we did at Hilbert's Hotel.

Now imagine a different scenario. You invite friends, and each one arrives with their own infinite set of friends—one has an infinite number of pals, another has an infinite family, and soon it turns into a swirling crowd of uncountable guests. You can't just expand your finite guest list to accommodate them; it's beyond what you originally planned. This is where the complexity of uncountable infinity truly stands out, revealing how vast and exciting these ideas can be.

Let's pause for a moment and reflect on how countable and uncountable infinity reflect our understanding of reality. We often find comfort in what we can see and measure. Yet, exploring these mathematical concepts nudges us to think beyond what's immediately in front of us. It reminds us that our understanding, like the natural numbers, is just a small piece of something much bigger— a peek into the infinite potential surrounding us.

This connection between counting and the unfathomable can also inspire deeper thoughts. What if we take this conversation beyond math and apply it to life and consciousness? Each unique perspective we have—our individual experiences, emotions, and dreams—can be compared to the natural numbers. They are countable, identifiable, and distinct. But when we try to capture the richness of human experience—our joys, heartbreaks, aspirations, and stories—we find ourselves facing an uncountable expanse that mere numbers can't fully describe.

As we explore these ideas, let's engage with thought experiments that push our thinking. For example, imagine trying to arrange all the possible outcomes of tossing a coin: heads or tails. If we only look at tossing the coin a few times—let's say three—we can easily list the results: HHH, HHT, HTH, HTT, THH, THT, TTH, TTT. But what if we think about tossing the coin an infinite number of times? Suddenly, we're confronted with an uncountable maze of possibilities. Each sequence represents a unique twist of fate, escaping the confines of our simple counting.

The beauty of exploring countable and uncountable infinity lies not just in their mathematical meanings but also in how they inspire our imaginations. They challenge us to push our limits and rethink the boundaries we sometimes unknowingly set for ourselves. Infinity isn't just a concept for mathematicians; it's a profound idea that can reshape our understanding of existence and the universe as a whole.

As we navigate the differences between countable and uncountable infinities, the world of mathematics unfolds as an exciting journey of thought. It encourages us to rethink what we know about size and limits, inviting us to dive deep into the complexities and embrace the endless possibilities that lie beyond our finite experiences. By engaging with these concepts, we open ourselves up to a realm where numbers dance in infinite harmony, patiently waiting for us to join the symphony.

Mind-Bending Implications

The idea of infinity is like a powerful lens that can change how we see the world. Yet, it often takes us on winding paths filled with paradoxes, confusion, and even moments of clarity. Consider Hilbert's Hotel, a thought experiment that we've touched on before. Picture a hotel that can welcome an endless number of guests, even when it seems completely full. This notion not only shakes our understanding of space and capacity but also makes us think deeply about what infinity really means. What happens when we try to take these intriguing ideas out of the realm of thought experiments and apply them to philosophy, mathematics, and even theoretical physics? The ideas that spring from this exploration can truly twist our minds.

Imagine a scenario where every time you believe you've wrapped your head around infinity, it slips away like sand through your fingers. Each attempt to define it feels like trying to grasp smoke with your hands. Infinity isn't just an impressive idea; it's a slippery puzzle that plays tricks on our understanding of reality. When mathematicians discuss infinite sets—whether they're countable or uncountable—they're not just talking about numbers; they're diving into the very essence of existence itself.

Let's take a moment to think about the philosophical implications of infinite sets and the puzzles they create. One of the most fascinating parts of infinity is how it pushes us to confront the limits of what we can understand. In our everyday lives, things are straightforward: if you have five apples, you know exactly what that means. But when we enter the world of infinity, everything changes. We start asking questions that can feel almost ridiculous: What does it really mean for something to be infinite? How can we wrap our minds around a quantity that has no end?

These questions lead us to even deeper philosophical discussions. What if the universe itself is infinite? Does that mean it contains countless versions of us, living out every possible twist and turn of our lives? If every choice we make creates a new branch of reality in an infinite multiverse, how does that change how we see ourselves and our sense of free will? The enormousness of infinity can leave us wrestling with existential questions, pondering our place in a universe that seems limitless and impossible to fully grasp.

Now, let's explore how infinity intertwines with mathematics and logic. The paradoxes that arise when we try to put limits on infinity can be dizzying. Take the Banach-Tarski Paradox, for example. It suggests that, in theory, you can take a solid ball, cut it into a finite number of pieces, and then reassemble those pieces into two identical solid balls. This mind-boggling idea comes from the unusual properties of infinite sets. It shakes up our fundamental understanding of volume and matter. How can something be divided and yet doubled? It's a reminder that our intuitive grasp of the physical world often fails when we step into the realm of infinity.

When we shift our focus to theoretical physics, the implications of infinity become even more striking. Concepts like black holes, singularities, and the vastness of space-time are all tangled up in infinite complexities. Take black holes as an example. At their centers lies a singularity, a point where gravity squashes matter to infinite density. At this point, our current understanding of physics completely unravels. Time and space lose their usual meanings, leaving us to wonder if we could ever truly understand what lies beyond the event horizon. What does it mean for our universe if it contains something—an area of infinite density—where our known laws of physics become meaningless?

Infinity also reaches into areas like quantum mechanics, where particles can exist in superpositions, acting both as particles and waves at the same time. This strange behavior hints at a deeper reality that is not only infinitely full of possibilities but also defies our traditional understandings of existence. As we dig deeper into the quantum world, we encounter questions that challenge our very ideas about certainty and determinism.

The implications of infinite mathematics extend beyond science and philosophy; they seep into our everyday lives. When you think of infinity, you might picture a vast expanse, like a never-ending desert. But what if infinity also touches the human experience? Consider emotions. Love, grief, happiness, and sorrow can be felt in countless shades, colors, and intensities. Each person's experience is unique, yet every emotion can be mapped to an infinite spectrum. In this sense, infinity becomes a tool for understanding the depths of human feelings, suggesting that our experiences can never truly be contained or categorized.

Now, let's shift our focus to the paradoxes that arise when exploring infinity. One well-known example is the Barber Paradox, which presents a scenario where a barber shaves all and only those men who do not shave themselves. The question then pops up: Does the barber shave himself? If he does, then he must not shave himself according to the premise. But if he doesn't, he belongs to the group of men who are shaved by the barber. This self-referential paradox highlights the complexities and contradictions that emerge when we try to apply finite definitions to infinite concepts.

Another brain-twisting example comes from Zeno's Paradoxes, which challenge our understanding of motion and division. Think about the famous Achilles and the Tortoise paradox. In this scenario, Achilles races a tortoise that has a head start. Zeno claims that Achilles will never catch up because every time he reaches the spot where the tortoise was, the tortoise has moved ahead a little more. This leads us to the conclusion that motion is impossible, even though we know from experience that it is not. These paradoxes push us to reflect on the nature of reality and the boundaries of our perceptions.

As we wrestle with these paradoxes, we're encouraged to rethink how we approach knowledge itself. Are we always held back by our limited understanding, or can we expand our perspectives and embrace the complexities of infinity? The beauty of exploring these ideas lies in their ability to stretch our imaginations and challenge our assumptions about what is possible. It sparks a sense of wonder and curiosity—qualities that are crucial in any intellectual endeavor.

Next, let's consider how these infinite concepts impact technology and our world. The computing power of today, especially in artificial intelligence, showcases the practical applications of understanding infinity. By using infinite algorithms and data sets, we are beginning to unlock new possibilities in fields like medicine and art. Picture a world where algorithms can analyze infinite combinations of genetic data to create personalized treatments for diseases, or where AI can produce art that feels endlessly varied, reflecting the unique experiences of each individual.

Understanding infinity can also inspire innovation and creativity. When we embrace the idea of infinite possibilities, we break free from conventional thinking. It encourages us to dream bigger and push the boundaries of what we think is achievable. Whether it's inventing new technology or imagining fresh forms of art, realizing that there are infinite ways to tackle a problem opens doors to ideas we may have never considered.

This idea of infinite potential can be empowering. Think back to the countless outcomes of a tossed coin. Each flip is a chance to explore a new path in life, a fresh decision, or an unexpected opportunity. It encourages us to see our choices not as fixed points but as part of an ongoing journey filled with limitless possibilities.

As we reflect on the implications of infinity, we shouldn't ignore the ethical questions that arise. Infinite resources, infinite knowledge, and infinite power come with serious moral considerations. How do we navigate a world where technology can create infinite copies of data, which raises issues of privacy and ownership? As we continue to innovate, it becomes crucial to think about the ethical implications of responsibly harnessing infinite possibilities.

Exploring infinity isn't just an academic exercise; it's a journey into the core of existence itself. It challenges our understanding of reality, pushes us to confront paradoxes, and invites us to rethink our connection with the universe. The beauty is in its complexity, where each revelation leads to more questions than answers. As we traverse this infinite landscape, we find ourselves continually engaged in a dance of thought, curiosity, and wonder—an endless quest to understand the ungraspable.

In this grand exploration, whether we're pondering the nature of reality, the limits of our knowledge, or the infinite possibilities of human experience, we are participating in a fundamental quest to understand not just infinity, but ourselves. As we navigate this vast, uncharted territory, we're reminded that in the search for knowledge, the journey itself holds as much significance as the destination. And perhaps, in recognizing this, we discover a small piece of the infinite woven into our very lives.

Avery Pascal

Chapter 2: Cantor's Diagonal Argument – When Infinity Comes in Sizes

The Infinite Landscape: Introduction to Cantor's Revolutionary Ideas

Imagine standing at the edge of an endless horizon—a vast ocean that seems to stretch on forever, a sky that knows no limits, a universe that goes beyond anything we can truly grasp. The idea of infinity has captured the imagination of philosophers, mathematicians, and curious thinkers for centuries. It's a word filled with wonder and mystery, challenging everything we believe about reality. In this immense expanse of thought, one name stands out, forever remembered in the history of mathematics: Georg Cantor.

Born in 1845 in the quiet town of Saint Petersburg, Russia, Cantor would change the way we think about infinity, introducing groundbreaking ideas that opened the door to a new era in mathematics. His journey, however, was far from easy. He faced considerable pushback from a mathematical community that clung to traditional views. Many respected mathematicians of his time found Cantor's ideas hard to accept, even outrageous. They were comfortable in the neat world of finite numbers and saw infinity as just a concept—more philosophical than mathematical.

But Cantor didn't let this opposition stop him. He envisioned infinity not as a single, unchangeable idea but as a rich and varied landscape. To him, infinity was like a world with different types of terrain—some areas vast and sprawling, others intricate and complex. To convey this groundbreaking idea, he suggested that not all infinities are the same. This notion became a cornerstone of his work and set the stage for his famous diagonal argument, which would demonstrate how infinities can differ in size.

Picture a packed stadium on a Saturday afternoon, with fans cheering for their team. This lively scene represents countable infinity; every person in the stands can be assigned a unique number, from the first fan to the last, no matter how far that last row stretches. Now, think about the endless ocean under a starry sky—each wave crashing ashore, each drop of water a unique entity. This endless stretch symbolizes uncountable infinity, a concept that can't simply be numbered or fully understood.

Cantor took these abstract ideas and turned them into a solid mathematical framework. He introduced the concept of cardinality, a way to measure the size of sets and determine if they are countable or uncountable. Imagine a library containing every book that could ever be written. This library represents uncountable infinity because the number of possible books is far greater than the countable infinity of whole numbers. Cantor's diagonal argument beautifully illustrates this, showing that even if you tried to list all possible real numbers, you'd always discover another that was missing from your list.

By introducing these ideas, Cantor not only reshaped our understanding of infinity but also laid the foundation for set theory. His work influenced many fields, from mathematics to philosophy, and even reached into computer science and logic. It's safe to say that Cantor's contributions changed the course of mathematics and how we view the infinite.

Yet, as he made strides in his work, Cantor faced intense backlash. Many of his peers considered his ideas radical, even heretical. Their resentment was more than just a disagreement over ideas; it turned into a personal struggle for Cantor, set against a rapidly evolving mathematical backdrop. He endured harsh criticism, and at times it pushed him to the edge of despair. Still, he found comfort in his firm belief that the truth of his findings would eventually shine through.

Cantor's ability to persevere serves as a powerful reminder of the strength of intellectual bravery. His story is not just about the mathematics of infinity; it's also about the resilience of the human spirit—its capacity for creativity, innovation, and, ultimately, triumphing over adversity.

As we delve deeper into the fascinating world of Cantor's ideas, we'll uncover how his work dramatically changed our perspective on mathematical thought. The idea that different infinities exist, each with its own unique traits and characteristics, invites us to think in a completely new way about numbers, sets, and the nature of existence itself. Through Cantor's groundbreaking insights, we'll explore the rich landscape of infinity, navigating its paradoxes and contradictions with a sense of awe and curiosity.

Our journey through Cantor's diagonal argument will reveal just how deeply these concepts resonate, encouraging us to rethink our assumptions about size, quantity, and the infinite. So, take a moment to open your mind to the possibilities that lie ahead. Embrace the beauty of this infinite landscape, and let's set off together into the captivating world of Georg Cantor's mathematics, where infinity comes in various sizes, and the adventure is just beginning.

Counting the Uncountable: Understanding Uncountable Sets

We've all faced the challenge of trying to count things that seem endless. Maybe it was a jar of jellybeans with flavors that tickle your taste buds, or perhaps it was the stars twinkling overhead on a clear night. In those moments, we come face to face with our limits, struggling to wrap our minds around the vastness of what we see. In the fascinating world of mathematics, there's a surprising difference between things we can count, like jellybeans and stars, and those that escape our counting efforts, such as the set of real numbers.

Let's take a moment to think about natural numbers. Imagine a long line of kids waiting eagerly for an ice cream truck on a hot summer day. Each child gets a number, starting from one and stretching on forever— two, three, four, and so forth, all the way to infinity. This idea is what we call countable infinity. Isn't it comforting? No matter how far you go in counting, there's always another number right around the corner, making it feel like a never-ending adventure. Importantly, we can list these numbers in order, creating a clear and logical sequence. Each natural number can be written down one after another, like an endless parade of digits that seem to stretch beyond what we can see.

But as we step beyond this familiar path, we run into a more complicated scenario: the set of real numbers. Here, the counting game takes an unexpected twist. Picture yourself at a lively party filled with laughter and music. People are chatting, sharing stories, and as is often the case, someone suggests a game: let's list all the real numbers! Sounds easy, right? But as soon as the idea is put out there, a chill runs down your spine. How could you possibly list every real number?

To understand this better, let's dive into Cantor's diagonal argument, a clever idea that shows why the set of real numbers is uncountable. Imagine you're at that party, surrounded by a crowd of enthusiastic participants. You all gather to make your list and scribble down some real numbers in decimal form:

1. 0.123456789...
2. 0.987654321...
3. 0.111111111...
4. 0.314159265...
5. 0.271828182...

As you keep going, the atmosphere is buzzing with excitement. You might think, "Of course, with enough time, we can list them all!" But here comes the twist. Cantor's diagonal argument tells us that for any list you make, there's always a way to create a new real number that isn't included in your list. How does this work?

Take the first digit from the first number, the second digit from the second number, the third digit from the third number, and so on. Then, simply change each digit. For instance, if the first digit is a '1,' switch it to a '2,' and if it's a '0,' change it to a '1.' This creates a new number that differs from every number in your original list by at least one digit. If your list was truly complete, how could this new number have slipped through? Yet it does, highlighting the brilliance of Cantor's idea: no matter how hard you try to count all real numbers, there will always be a number that remains unlisted.

This discovery is revolutionary. It shakes up our understanding of infinity itself. If we can count the natural numbers, why can't we count the real numbers? Aren't all real numbers just different points on the number line? The answer lies in the nature of infinity.

Let's think of the difference between a crowd at a concert and an ocean of waves. The natural numbers are like that concert crowd; they can be counted, and everyone can get their own distinct number. On the other hand, real numbers are more like the ocean—an ungraspable expanse filled with countless waves, each representing a unique real number. Plus, real numbers include not just whole numbers but fractions, and irrational numbers like π and $\sqrt{2}$, and everything in between. There are so many real numbers that between any two of them, no matter how close, there are infinitely more in between.

Now, let's think about what this amazing difference means. This isn't just a dry concept from a math textbook; it changes how we see size and comparison. It forces us to face the fact that not all infinities are the same. While the natural numbers are infinite, they are just a tiny piece of the mind-blowingly vast world of real numbers.

As we navigate these mathematical waters, the idea of uncountability reshapes our understanding of both math and reality. It makes us rethink our ideas about size, quantity, and what infinity really means. In Cantor's world, infinity isn't a single idea; it's a diverse landscape filled with complexities that invite us to explore further.

And so, as our party continues, filled with laughter and spirited discussions about numbers, we come to realize that the joy of mathematics isn't just in the answers we find;

it's in the questions we ask. It's about the journey of exploration, the thrill of discovery, and the mind-bending revelations that unfold as we dig deeper into the world of math.

In this lively and thought-provoking atmosphere, we can see how Cantor's diagonal argument opens doors for understanding not just infinity, but also the essence of reality itself. It encourages us to embrace the puzzles that come with the infinite and challenges our traditional ideas about existence. Standing at the edge of this mathematical ocean, ready to dive in, we learn that the adventure is just getting started.

Ultimately, Georg Cantor's legacy is more than just the math of infinity. It's a celebration of curiosity, a tribute to those who challenge the norms, and an invitation to anyone willing to embrace the complexities of the infinite. As we continue to explore these ideas, it's clear we'll be captivated by the beauty, mystery, and endless possibilities that lie just beyond the horizon of our understanding.

Beyond the Infinite: The Impact of Cantor's Work on Mathematics

Many people see mathematics as a chilly, unfeeling space—a place where numbers and symbols swirl around in textbooks, giving life to ideas that can sometimes feel remote. However, beneath this seemingly stark surface lies a rich blend of human thought, creativity, and, at times, heated debate. At the center of this mathematical adventure is Georg Cantor, a man whose groundbreaking ideas about infinity sparked both admiration and criticism in equal measure. Cantor's work has left a lasting legacy, reshaping the landscape of modern mathematics in ways we continue to explore today.

Cantor laid the foundation for what we now know as set theory, a crucial building block in today's mathematics. To fully appreciate how monumental this was, it's important to understand the challenges he faced in getting the math community to embrace his ideas. When Cantor first proposed that there were different sizes of infinity, he wasn't just tweaking the rules; he was challenging how mathematicians viewed numbers and their relationships.

Picture a room full of mathematicians in the late 19th century, a time when most people barely talked about infinity. In a world where there was a common belief in a single, unchangeable concept of infinity, Cantor stepped forward with a bold claim: infinity comes in various sizes, and some infinities are actually bigger than others. For Cantor, this wasn't just fanciful thinking; it was a well-reasoned argument supported by solid proofs.

Yet, as many historians point out, new ideas often face resistance, and Cantor's were no different.

Criticism rolled in quickly from respected figures like Leopold Kronecker, who famously dismissed Cantor's work as "a mere figment of the imagination." For Kronecker and his allies, Cantor's ideas were an attack on the established norms of mathematics, a discipline that valued the clear and the finite. They viewed Cantor's exploration of infinity as unnecessary and potentially chaotic. This pattern of skepticism is common in the history of science and mathematics, where groundbreaking theories often encounter backlash from those who hold the traditional views.

But for every critic, there were supporters who appreciated Cantor's genius. Slowly but surely, mathematicians began to embrace his ideas, leading to an explosion of research that deepened our understanding of infinity. Cantor's work on cardinal numbers and transfinite ordinals allowed mathematicians to explore the vast realms of infinite sets with newfound confidence and clarity.

One of the most significant developments stemming from Cantor's theories was in the field of topology, which looks at the properties of space that remain unchanged under continuous transformations. Cantor's ideas about the different sizes of infinity sparked discussions about continuity, compactness, and convergence. Mathematicians like Henri Poincaré and David Hilbert began to adopt Cantor's frameworks, applying them to questions about space and dimensions that had been largely ignored. Instead of seeing infinity as a frightening abyss, they began to recognize it as a fascinating landscape waiting to be explored.

Cantor's influence also significantly shaped the development of mathematical logic. The ideas he introduced about functions, relations, and cardinality became a foundation for later thinkers like Kurt Gödel and Bertrand Russell, who would re-examine the very foundations of mathematics. Cantor's insights into infinite sets provided a basis for tackling paradoxes and inconsistencies that had long troubled the field.

However, Cantor's journey was not without its struggles. The harsh criticism he faced took an emotional toll; he experienced bouts of depression due to the rejection of his ideas by peers. His letters reveal a man wrestling with not just mathematical concepts but also the pain of isolation. This highlights a poignant truth about great thinkers: their monumental ideas often come at a personal cost. Despite the challenges, Cantor found comfort in his conviction about the significance of his work, famously stating, "The essence of mathematics lies in its freedom." This reflects the unyielding spirit of inquiry that fuels all great mathematicians.

Looking back on Cantor's legacy prompts us to consider the wider implications of his work. The exploration of infinity has seeped into areas beyond pure mathematics, influencing philosophy, physics, and even the arts. In philosophy, discussions about infinity challenge our understanding of existence and reality itself. Philosophers have long debated whether we can truly grasp infinity or if it's just a concept, a conversation that continues to resonate today.

In physics, the ideas about infinity push us to rethink our understanding of the universe. The concept of an infinite universe raises big questions about the very fabric of space and time. As cosmologists explore the edges of what we can observe, they grapple with the implications of infinite space, time, and matter. Cantor's insights urge us to confront our own limitations as we try to make sense of the infinite alongside the finite.

As we engage with these profound ideas, we are reminded that infinity isn't just an abstract concept limited to math textbooks. It seeps into our daily lives, encouraging us to reflect on the endless possibilities that lie ahead. Consider the endless cycle of questions that arise when we ponder existence itself. Our attempts to understand infinity echo our search for meaning in a world that often feels overwhelming.

Cantor's work encourages us to welcome the complexity of infinity and to explore the mysteries within this vast universe. It challenges us to think critically, not just about numbers, but about the essence of reality itself. Just as Cantor navigated the often-choppy waters of mathematical thought, we too are invited to dive deep into our curiosity, to ask questions that might not yield simple answers.

In a world that often craves certainty and clear-cut resolutions, embracing complexity can be a bold choice. Cantor's legacy is a powerful reminder that progress often comes with hurdles and that some of the most profound discoveries emerge from our willingness to engage with uncertainty. The mathematicians who followed Cantor have shown us that the journey of exploration is just as valuable as reaching a destination.

Let us hold tight to Cantor's pioneering spirit as we contemplate the infinite possibilities that lie ahead. Embracing the complexity of infinity opens up new avenues of understanding, inviting us to appreciate the beauty, wonder, and interconnectedness of everything. As we marvel at the mysteries of the universe, we realize that the adventure of inquiry is endless—much like the concept of infinity itself.

In a world that sometimes feels restricted by the limits of finite understanding, we can look to Cantor's work as an invitation to broaden our horizons. The exploration of infinity goes beyond the borders of mathematics; it becomes a philosophical journey, an artistic adventure, and a scientific pursuit. It encourages us to welcome the unknown, to engage with the mysteries of existence, and to keep asking the questions that stretch our understanding.

As we navigate the complexities of our lives, let's carry the lessons learned from Cantor's journey with us, celebrating the spirit of curiosity and the quest for knowledge that defines our humanity. In doing so, we honor not just Georg Cantor's legacy but also the endless possibilities awaiting those brave enough to venture beyond the known and into the infinite.

Avery Pascal

Chapter 3: Zeno's Paradoxes – The Race That Never Ends

The Dichotomy Paradox

Picture this: you're gearing up for a casual walk to your local store, or maybe you're feeling competitive and decide to race your friends, the kind of friendly challenge that brings back good memories. The excitement builds as you take that first step. But just before you reach that store or cross the finish line, a little voice pops into your head, reminding you of an ancient philosopher named Zeno. According to Zeno's Dichotomy Paradox, before you can reach your goal, you have to cover half the distance first. But wait—before you can even get to that halfway point, you need to cover half of that distance. And this keeps going on forever.

It's a bit of a head-scratcher, isn't it? If you keep breaking down each step you take into tinier and tinier fractions, can you really ever make it? Just when you think you've reached halfway, there's another halfway point waiting for you. This paradox seems to suggest that motion itself is just an illusion, a trick the mind plays, leaving you caught in this endless cycle of "almost there."

To really grasp Zeno's point, let's think about that journey to the store. You set off, feeling pretty good about it. You make it through the first half of the distance—let's say the store is 100 meters away, so you've walked 50 meters. But then, as Zeno would say, before you can tackle that last 50 meters, you need to get to the halfway mark, which is 25 meters. Easy, right? But then, before you can cross that 25-meter line, you have to walk 12.5 meters. And just when you think you're close to the store, the distances shrink down to 6.25 meters, then 3.125 meters, and on it goes.

You might find yourself chuckling at the silliness of this thought experiment. Picture someone so wrapped up in walking that they get stuck in this ridiculous loop of calculations, never quite making it to the store but always getting closer. It's a funny image, maybe of a friend, determined but hilariously trapped in a math puzzle.

This paradox, introduced by Zeno over two thousand years ago, wasn't just a random thought; it posed a serious challenge to our understanding of motion and existence. Zeno lived in a time buzzing with philosophical exploration. His ideas struck a chord with the ancient Greeks, who were deeply interested in the nature of reality and what it means to exist. Philosophers back then were striving to understand the universe around them while wrestling with their own ideas. Does motion even exist if it requires an endless number of steps to achieve? Are we just spectators in a world where we can never really travel the distance between two points?

Throughout history, philosophers like Aristotle tried to counter Zeno's arguments. They believed that while reaching a destination might involve infinite steps, the idea of motion itself is still valid. Aristotle introduced the idea of actual infinity versus potential infinity, arguing that although we can think about dividing distances infinitely, we can also act on those divisions in the real world. Yet, Zeno's paradox sparked discussions that have echoed through time, inspiring debates that still resonate in modern math.

To make Zeno's paradox even clearer, consider the classic race between a tortoise and a hare. In this famous tale, the hare, full of confidence, lets the tortoise start first. As the race kicks off, the hare dashes ahead, covering the initial distance. But Zeno would argue that every time the hare runs a distance, it has to cover half of that distance first, and then half of that remaining distance, and so on. By the time the hare reaches the tortoise's starting point, the tortoise has moved a bit further ahead. This creates a seemingly endless loop—so can the hare ever really catch up?

The humor and absurdity of this race lie not in who wins, but in how it makes us rethink our concepts of motion and time. Imagine the scene where the hare, filled with bravado, tries to gain ground but ends up watching the tortoise inch forward, all while he breaks the race into tiny segments. It's a funny take on competition, where skill gets overshadowed by the complexities of infinity.

As you think more about this paradox, consider your own life. Have you ever faced a task that seemed impossible? Maybe a big project at work or a personal goal—each step felt like it would take forever to complete. When we get too caught up in the details, we can feel stuck, trapped in our own versions of Zeno's paradox.

But why should this matter to you? Why should Zeno's tricky logic and philosophical dilemmas catch your attention? These paradoxes are fascinating not just for their complexity, but for how they challenge us to think about our assumptions regarding reality. They push us to question our views on time, space, and what it means to take action. In our busy lives filled with distractions and a craving for instant results, Zeno reminds us that sometimes, the journey itself can feel endless.

This is where Zeno's Dichotomy Paradox opens up a doorway to critical thinking and philosophical exploration. It invites us to engage with our thoughts, challenge what we take for granted, and reflect on our understanding of existence. Each time we undertake what seems like a simple journey, we can pause and acknowledge the infinite steps that stand between us and where we want to go.

As we ponder the deeper meanings of motion—or the lack of it—Zeno's paradox leaves us questioning not just the nature of reality but also our place in it. While motion may be an illusion, the ideas it sparks are undeniably real. So, even if you feel the urge to lace up your sneakers and dash to the store, remember that every journey, no matter how small, is filled with the intriguing complexities of infinity.

Infinite Series Summed

As we explore the fascinating world of Zeno's paradoxes, where the idea of motion often feels like a riddle, we find ourselves at an intersection where mathematics meets philosophy in surprising ways. The Dichotomy Paradox invites us to think about how, before we can take a single step, we must first cover an infinite number of smaller distances—each time halving the space we have left to travel. While Zeno's ideas might seem confusing, the beauty of mathematics, especially through calculus, shows us that these endless divisions can actually lead to clear and finite results.

Let's start with a key idea: the limit. In calculus, the limit helps us understand how functions behave as they get closer to a certain point. Think of it like watching your favorite TV show. Instead of waiting for the last episode, you get to see how the story builds towards an exciting climax. Just like a story reaching its end, a mathematical function approaches its limit, giving us insights into motion, speed, and distance—even when we wrestle with Zeno's paradoxes.

To make this idea clearer, imagine a simple geometric series, which nicely shows how things that seem infinite can actually result in something finite. Picture yourself with a pizza, a treat many of us adore. You start by slicing the pizza in half. Then, you take one of those halves and cut it in half again. You keep repeating this process endlessly. Your first piece is $1/2$ of the pizza, the second piece is $1/4$, the third is $1/8$, and it just keeps going. If we add up all these tiny pieces, we find that they total a whole pizza:

$$1/2 + 1/4 + 1/8 + 1/16 + 1/32 + ...$$

This is the magic of an infinite series. Using the formula for the sum of an infinite geometric series, $S = a / (1 - r)$, where 'a' is

the first term and 'r' is the common ratio, we can find out how much pizza we actually have. Here, 'a' is $1/2$, and 'r' is also $1/2$. If we plug these values into the formula, we get:
$$S = (1/2) / (1 - 1/2) = (1/2) / (1/2) = 1.$$

Amazing, right? All those infinitely small pieces of pizza equal one whole pizza. Isn't it delightful to think that even as you keep slicing the pizza forever, you still end up with a full and delicious pie?

The work of mathematicians like Archimedes, who was thinking about these ideas way before calculus came along, set the stage for our modern understanding. Archimedes had a remarkable way of approaching areas and volumes, using methods similar to limits, which helped him accurately calculate the areas of circles and other shapes. He showed that the area of a circle can be understood as the limit of the areas of inscribed polygons as the number of sides increases. If Archimedes could have seen the advancements made by calculus later on, he would have been thrilled by the clarity we now have around infinitesimals.

Fast forward to the 17th century, when calculus—thanks to thinkers like Isaac Newton and Gottfried Wilhelm Leibniz—changed the way we see motion and change. They gave us new ways to think about infinite series. The derivative, which helps us find instantaneous rates of change, and the integral, which lets us sum up countless tiny quantities, changed the game in mathematics.

But let's not get too caught up in the fame of historical figures just yet. Let's bring it back to reality. Imagine being at a racetrack with the sun shining down, the smell of popcorn in the air, and the crowd buzzing with excitement. The racers are on the track, and as the starting gun fires, each one takes off. They cover not just one distance but countless smaller segments within that distance. If we look closely at their movements, we see that they, too, are participating in this intricate dance of infinity. Every step they take can be thought of as an infinite series of movements, yet they still manage to cross the finish line. How is that possible?

This is where the beauty of calculus shines through. Each time a racer moves forward, they seem to be part of an infinite sum of tiny distances, but they still reach the finish line in a reasonable amount of time. This endless series of steps allows them to achieve their goal.

Now, let's break it down with some numbers and see what it looks like on paper. Picture a runner starting at the starting line (0 meters) who wants to reach the finish line at

100 meters. If they run in a way that continuously halves the remaining distance, they might first cover 50 meters, then 25 meters, then 12.5 meters, and so on. As we look at this series, we can express their movements mathematically:

$$S = 50 + 25 + 12.5 + 6.25 + ...$$

Using the geometric series formula again, we can start to see the pattern. Here, the first term (a) is 50 meters, and the common ratio (r) is 1/2. Plugging these numbers into the formula gives us:

$$S = 50 / (1 - 1/2) = 50 / (1/2) = 100.$$

In this case, the runner effectively covers the entire distance of 100 meters, despite taking an infinite series of steps along the way. This math helps us reconcile Zeno's paradox with the reality of motion, allowing us to appreciate the complexity of our universe without getting stuck in the logic of infinity.

Another fascinating aspect of this discussion is the idea of convergence. In math, convergence refers to the way a sequence approaches a specific value as it continues on towards infinity. Imagine it as a journey where, even though you face endless challenges, you are steadily heading toward your destination. The limitations posed by Zeno's paradox become less intimidating when we grasp the idea that some infinite sums can converge to a finite value.

Visualizing these concepts can greatly aid our understanding. Graphs and diagrams help us see the results of infinite series and limits in a more concrete way. Picture a graph where you can observe a curve that approaches a line without ever quite touching it. This behavior, known as asymptotic behavior, reveals so much about the relationship between motion and infinity, inviting us to visualize how we can strive toward a goal even if we never fully reach it—an idea that resonates in both math and life.

Engaging with these mathematical ideas can be an enriching experience. As we understand limits and infinite series, we aren't just spectators; we become active participants in the amazing world of mathematics. Try calculating an infinite series on your own. Using the geometric series we discussed, take a moment to sum different series with various starting points and ratios. You might find yourself enchanted by the elegance of these numbers.

Maybe you have a favorite series that leads to a surprising outcome, or you can think of a real-life situation that illustrates the ideas of convergence and limits. Let your creativity flow! This exploration of math allows us to appreciate the beauty and complexity of our universe while recognizing that navigating the infinite can lead us to clear, meaningful conclusions.

As we keep exploring the vast world of mathematics, Zeno's paradox reminds us that our understanding of reality is always growing. The mix of philosophy and math isn't just an academic exercise; it's a way to gain deeper insights into existence, time, and our understanding of reality. Even as we wrestle with the complexities of motion and infinity, we find comfort in knowing that we can travel through this infinite landscape, reaching tangible conclusions and savoring the rich exchange of ideas that math offers.

So, whether you find yourself racing against time, pondering the wonders of an infinite series, or simply enjoying a slice of pizza, remember that the quest for understanding is a journey filled with limitless possibilities. And while Zeno challenged our ideas about motion, we have the tools to work through these puzzles, not just as mathematicians but as curious explorers who dare to question the very fabric of existence. As we ponder the infinite, we come to see that every limit we approach, every series we sum, is a bridge that brings us closer to the essence of understanding.

Bridging Philosophy and Math

Picture a lively café on a cool autumn morning, filled with the rich smell of fresh coffee and the gentle buzz of chatter. At a cozy corner table, two people are deep in conversation, bouncing between big ideas and everyday realities. Clara is a passionate mathematician with a love for numbers—and coffee. Sitting across from her is Victor, a philosopher who has a talent for turning complicated thoughts into engaging stories. Their exchange captures the lively connection between mathematics and philosophy, bringing to life the themes sparked by Zeno's paradoxes.

"Zeno's paradoxes," Clara says, stirring her coffee with a thoughtful look, "are a clever way to show how tricky motion can be. They ask questions that really make us think about space and time. Take the Dichotomy Paradox, for example. Before you can reach your destination, you first have to cover half the distance. But then, to cover that half, you need to cover half of that half, and this goes on forever. It leaves us with an endless number of steps to finish a single journey."

Victor leans in, his interest piqued. "So, Zeno is basically suggesting that movement is impossible because it involves completing an infinite number of steps in a limited amount of time. But what does that say about our view of reality? Should we believe that the physical world clashes with mathematical reasoning?"

Clara laughs lightly, excitement shining in her eyes. "That's where modern math shines! Calculus helps us tackle this paradox. The idea of limits allows us to add up those infinite steps into a finite result. It may seem like we take an endless number of steps, but mathematically, we can get close to a limit that gives us a clear answer. Think about a runner starting at zero meters and aiming for 100 meters. By always halving the remaining distance, they seem to take infinite steps, but they still cross the finish line!"

Victor sits back, absorbing the implications. "It's amazing how calculus can turn what feels like an impossible philosophical question into a solvable math problem. But how does this connect to our understanding of existence? Does this mean that reality is more like a math concept than something we can actually observe?"

"Ah, that's the heart of the issue!" Clara replies, her enthusiasm contagious. "Philosophers like Leibniz and Kant have wrestled with the ideas of infinity and existence. Leibniz believed there are countless possible worlds, while Kant argued that our grasp of time and space is shaped by how we process our sensory experiences. So, in a way, Zeno's paradoxes have opened the door for deeper philosophical discussions about what it means to exist."

As they chat, their dialogue feels like a graceful dance, moving between math and philosophy. Clara adds, "The relationship between these two fields is a rich one. They challenge and support each other, pushing their ideas further. When we think about infinity, we're not just looking at a number that goes on forever. We're exploring a concept that shapes our understanding of the universe—its structure, its limits, and even its very essence. Thinkers from Aristotle to modern philosophers have tried to define infinity, often wrestling with its paradoxes."

Victor nods, his mind buzzing with thoughts. "This brings us to modern debates about space and time. Physicists are questioning whether time is continuous or made up of separate moments. If we can solve Zeno's paradoxes with calculus, could there also be a mathematical way to tackle these philosophical questions about time? Are we just starting to grasp the true nature of reality?"

Clara leans in, her eyes sparkling with interest. "Exactly! Consider this: physics has always been rooted in mathematical principles. Look at how the theory of relativity changed how we view time and space. Just as Zeno's paradoxes sparked reflections on infinity, they also influenced the development of ideas that challenge our reality. How do we reconcile the infinite divisibility of space with our finite physical experiences? That's a question both mathematicians and philosophers must face."

The café hums around them, but Clara and Victor are lost in their conversation, the busy world outside fading away. "I like to think of mathematics as a language," Clara muses. "It has its own rules and structure, just like any spoken language. But it's a language that helps us express ideas about infinity, continuity, and existence with clarity. When we talk about Zeno's paradoxes, we're not just playing with numbers; we're engaging with the very essence of reality and our role in it."

Victor takes a sip of his coffee, the steam swirling in the air. "And this connection between math and philosophy encourages us to reflect on our own views. What does it mean for something to be real if it can be described mathematically yet feels distant in our everyday experiences? Are we living in a mathematical framework, or is that just a tool we use to make sense of the chaos around us?"

Clara smiles knowingly, appreciating the depth of Victor's thoughts. "Both views are valid. Mathematics gives us clarity, helping us navigate the infinite complexities of existence. But philosophy enriches our understanding, urging us to think about the nuances and implications of those mathematical truths. It's a delicate dance that has fascinated thinkers for ages."

As their conversation unfolds, they touch on the ideas of philosophers like David Hilbert, who famously said, "The infinite is not just a concept; it's a reality." This idea pushes us to broaden our understanding of the universe beyond our limited experiences. It encourages us to embrace the endless possibilities that come from the blend of mathematics and philosophy.

"Consider Cantor's work on set theory," Clara suggests. "He introduced the idea that not all infinities are the same. This challenges our foundational ideas about mathematics and opens the door to a deeper understanding of reality. This is where philosophy and math intersect, as we explore how these concepts affect our view of the world."

Victor's eyes widen with curiosity. "So, as we wrestle with Zeno's paradoxes, we're not just solving math puzzles; we're diving into a deeper exploration of existence itself. The concept of infinity challenges us to face our limitations and broaden our understanding. It pushes us to ask: what is reality like when we confront the infinite?"

Clara nods, her excitement evident. "Exactly! These paradoxes remind us that our instinctive view of reality might not always match mathematical truths. They encourage us to question our assumptions and consider a wider range of possibilities. This meeting point of math and philosophy isn't just an academic exercise; it's a meaningful exploration of what it means to live in an infinite universe."

Their conversation flows effortlessly, filled with laughter and shared insights as the café bustles around them. As they delve into Zeno's paradoxes, they find themselves in a delightful paradox of their own: two distinct realms—math and philosophy—intertwined, shedding light on the complexities of existence.

The exchange between Clara and Victor reflects the ongoing dialogue between numbers and ideas. Each brings their unique viewpoint, yet they discover common ground in their quest for understanding. They embody the belief that tackling philosophical questions can lead to new insights in math, and that discoveries in math can inspire deeper philosophical inquiry.

As their coffee cups empty, they acknowledge the power of this intersection.

Their discussion about Zeno's paradoxes serves as a reminder that seeking knowledge is a shared adventure. It invites more exploration and reflection, encouraging us to ponder the infinite complexities of the universe and our place in it.

Eventually, Clara and Victor realize that the richness of their conversation will linger long after the last sip of coffee. The questions raised during their time together will continue to echo in their minds, prompting them to engage with the infinite and the finite in ways they have yet to explore.

As the sun begins to set outside the café, casting a warm glow through the windows, their conversation fades into the background noise. Yet, as they step out into the world, they carry with them the spark of curiosity ignited by their discussion—a reminder that the journey of understanding is never truly finished.

Mathematics and philosophy, much like the coffee they shared, blend together to create a rich mix of insights that challenge, inspire, and provoke thought. Just as Zeno's paradoxes encourage us to question our views on motion, they also push us to think about the very nature of infinity, existence, and the universe itself.

In this lively exchange, Clara and Victor exemplify the true spirit of connecting philosophy and math. Their discussion invites readers to explore the endless possibilities that arise when we engage with these profound ideas, reminding us that the journey of understanding is as boundless as the concepts it covers. With the universe as our canvas, we are left wondering: what lies beyond the limits of our knowledge, waiting to be uncovered?

Avery Pascal

Chapter 4: Thompson's Lamp – Flickering Forever in Finite Time

Supertasks Explained

In the quiet corners of academia, where the ticking clocks and chalk dust create a unique atmosphere, we encounter a fascinating idea that challenges how we think about time and space. Imagine a student sitting in a bare classroom, grappling with the confusing idea of infinity during a tough math lecture. The professor stands at the front, armed with a whiteboard filled with complex symbols, trying to explain the concept of tasks that extend beyond the limits of what we consider possible. Yet, the student struggles to wrap their head around how anyone could complete an infinite number of tasks in just a short amount of time. It's a mind-bending puzzle that encourages all of us to rethink what infinity really means.

This is where the exciting world of supertasks comes into play. Simply put, a supertask is when someone completes an infinite series of tasks in a limited amount of time. At first, this might sound like something out of a science fiction story, but it's actually grounded in solid mathematics. Picture this: a traveler needs to walk halfway to a destination, then halfway again, and then halfway once more, and so on, forever. It seems like they would never actually get to their destination, but the math behind this curious situation helps us see the idea of infinity in a whole new way.

One of the best examples of supertasks can be found in one of the most famous puzzles in math: Zeno's Paradoxes. Among these, the story of Achilles and the Tortoise stands out. In this thought experiment, the fast Achilles races against a slow tortoise that has a head start. One would expect Achilles to easily outrun the tortoise. However, Zeno's argument leads us into a maze of infinite steps, showing us that by the time Achilles reaches the place where the tortoise began, the tortoise has moved ahead, even if by just a little. This cycle continues forever. The paradox makes us consider the idea that while Achilles is faster, all the infinite steps he has to take make it seem like he can never catch up. It's a fascinating dance with infinity that leaves us thinking about the nature of time itself.

As we explore the deeper meanings behind supertasks, we encounter the remarkable work of mathematicians like Georg Cantor and David Hilbert, whose studies of infinity have greatly influenced our understanding of the infinite universe. Cantor introduced the groundbreaking idea that not all infinities are the same, creating concepts like countable and uncountable infinity. His work encourages us to view infinity not as a single idea, but as a spectrum filled with different possibilities, each with its own unique quirks. Hilbert, on the other hand, asked us to think of infinity not just as a number, but as a concept that stirs both wonder and confusion. His famous thought experiment, Hilbert's Hotel—an imaginary hotel with infinitely many rooms—delightfully shows how infinite tasks can fit into finite spaces.

    While we reflect on the implications of supertasks, we find ourselves at a crossroads between math and philosophy, where the rules we know seem to bend and twist. It raises the question: Can we really understand what it means to perform an infinite number of tasks? This question nudges us to rethink our views on time and space, leading us to an interesting analogy that captures the heart of supertasks. Think of a series of movie trailers that tease a glimpse of an exciting cinematic universe. Each trailer offers a taste, hinting at the greatness to come, yet the full movie always seems just out of reach. The trailers play one after another, each providing a small preview of what's ahead, but never quite delivering the full experience. This endless cycle of anticipation and moments that never fully arrive echoes the strange nature of supertasks, where infinite tasks unfold in a limited timeframe, creating a dizzying dance of completion that feels almost impossible.

In this realm of thought, the idea of achieving the infinite transforms from a mere mathematical puzzle into a deep philosophical question. Are we just players in a script written by the whims of infinity, or do we have the power to navigate through the endless tasks that surround us? As we peel back the layers of this paradox, we discover that exploring supertasks reveals insights not only about math but also about our perceptions of reality. The flickering light of Thompson's Lamp shines brightly, highlighting the complexities and contradictions we face in trying to grasp the infinite within the limits of time. Each flicker invites us to reconsider what we know and embrace the fascinating dance of infinity that plays out all around us.

As we think about these supertasks, we step into a wonderful space where logic unravels and creativity soars. The journey through infinity isn't just an abstract exercise; it reflects our daily lives and the countless tasks we manage. Just as that student in the math class wrestles with infinite sequences, we too confront the infinite nature of our own experiences. The flickering of Thompson's Lamp reminds us that within the infinite lies a paradox that may never be fully resolved, yet it continues to illuminate the mysteries of our universe and our place within it. This contemplation sparks curiosity, encouraging us to explore the endless possibilities waiting to be discovered.

As we dive deeper into the captivating world of supertasks, we are prompted to think not only about the math involved but also about the bigger questions that come from engaging with infinity. It invites us to reflect on our own experiences with time and space and how they connect with the rich narratives of human understanding. Will we rise to the challenge of grappling with infinity, or will we seek comfort in the familiar confines of the finite? As we navigate this intriguing landscape of supertasks, we open ourselves to a fuller understanding of both mathematics and what it means to be human in an infinite universe.

The On-Off Paradox

Picture a cozy little room, filled with charming knick-knacks that whisper tales of the past. In the center, casting a warm glow across the wooden floor, sits a delightful bedside lamp—its soft shade resting on a slender brass base. This isn't just any lamp; this is Thompson's Lamp. With a flick of the switch, this lamp takes us on a curious journey that challenges what we think we know about time, light, and infinity.

Now, imagine this lamp in action: the switch turns on, flooding the room with brightness, lighting up every corner and chasing shadows away. Then, just as quickly, the switch flips off again, plunging the room into darkness. On, off, on, off—it's a relentless cycle. Here's where it gets intriguing: this back-and-forth doesn't stretch out over days or even hours; it all happens within a single minute. That's right—within sixty seconds, this lamp will toggle an infinite number of times. And this leaves us with a puzzling question: at the end of that minute, is the lamp on or off?

Let's break it down a bit. The lamp starts by being on for the first half-second. Just when the light fills the room, it turns off for the next quarter-second. Then it bursts back on for an eighth of a second, only to shut off again for a sixteenth. This pattern goes on infinitely, with each interval getting half as long as the last. If you were to write this down, it would look like this: 1/2, 1/4, 1/8, 1/16, and so on. Each flick of the switch happens faster and faster, and even though there are infinitely many events, they all take place in just one minute.

As the first half of the minute ticks by, the lamp is on for countless tiny slices of time, seemingly lighting up the room. But we can't ignore the time it spends turned off. If you pause for a moment, you might wonder: how much time does it really light up the room compared to when it's in darkness? Those fractions of a second add up—a blur of toggling that leaves us guessing about the final state of the lamp as the minute wraps up.

This brings us into the realm of math, where we start to play with ideas of limits and convergence. At the end of the minute, we might want to make a snap judgment about whether the lamp is on or off, but math has a way of challenging our gut feelings. The heart of this paradox comes from questioning whether the infinite number of "on" states cancels out the infinite number of "off" states, or if one ends up being more dominant.

To make sense of this, let's look at the idea of limits. We can express the total time the lamp is on as a series of fractions:

$1/2 + 1/4 + 1/8 + 1/16 + \ldots$

Adding these fractions together shows us that as we keep going, the total approaches 1. In simpler terms, the lamp is on for just one second—the entire duration of that minute.

But what about those "off" states? We find a similar situation there. The lamp is off during these fractions of time:

1/4 + 1/8 + 1/16 + …

Surprisingly, this also adds up to 1 as we combine the fractions. So, while the lamp is on for one second, it also spends an equal amount of time turned off.

Now, our minds start to race: by the end of the minute, what's the final state of the lamp? It has switched infinitely between on and off, and yet it feels perfectly balanced! This is the paradox—how can the lamp be both on and off at the same time? It's a riddle that has confused many thinkers.

To make things even more interesting, let's introduce a thought experiment. Picture a close friend asking, "If I walked into the room right at the end of that minute, what would I see? Would the lamp be glowing softly, or would it be dark?" This question prompts us to consider how observation plays a role in determining what's real. If you were there at that exact moment, how would you know whether the lamp was on or off? It's a captivating idea that touches on quantum mechanics, where just looking at something can change the outcome.

Philosophers like Bertrand Russell and mathematicians like Kurt Gödel have wrestled with similar puzzles in their efforts to make sense of infinity. Their discussions often come back to how our understanding is shaped by the frameworks we use. In this case, the lamp serves as a metaphor for broader philosophical questions about existence and observation. Are things as straightforward as they seem, or do deeper complexities lie beneath?

As we ponder these ideas, we start to see that the implications of this lamp's paradox go beyond math. It nudges us to think about what we consider truth and certainty in a world that often lacks clear answers. The flickering of Thompson's Lamp becomes a symbol of the competing narratives in our search for knowledge: certainty versus doubt, light versus darkness.

Let's take a moment to think about how this concept appears in our daily lives. Consider the times when we find ourselves caught between choices. Should I take a leap into something new, or stick with what I know? In those moments, we may feel a mix of excitement and fear, just like the lamp flipping between light and the comforting shadows around it.

This serves as a reminder that life, like the lamp's paradox, isn't usually about absolute certainties. Each flicker of light in our lives intertwines with moments of darkness, creating a rich tapestry of experiences.

Returning to the lamp, the paradox it presents is a powerful invitation to rethink our understanding of infinity. It encourages us to explore the limits of our knowledge and recognize that what seems straightforward might actually be filled with uncertainty. By grappling with this paradox, we realize our grasp on reality is often nuanced and complex.

As we step back from this intricate discussion, the question lingers: is the lamp on or off? Perhaps the answer isn't as vital as the exploration this journey offers. Like the infinite flickers of Thompson's Lamp, our quest for understanding may illuminate just as much as the answers we seek.

This paradox invites us to embrace the beauty of uncertainty, to appreciate the complexity of life, and to treasure the moments that dance between light and dark. In a world where absolutes often rule, it is this dance of possibilities that keeps our minds alive, igniting our curiosity and wonder about the endless nature of existence. So, the next time you find yourself in a softly lit room, with a lamp flickering nearby, remember: within that simple act lies a profound exploration of reality, uncertainty, and all that lies in between.

Infinity Meets Reality

Imagine this: a vast cosmic stage, where stars sparkle like a billion tiny lights, twinkling on and off in a grand performance.

Infinity Paradoxes

Each celestial body isn't just a distant sun but a character in an epic play that stretches the limits of reality. Here, time isn't just a straight line from start to finish; it's a swirling experience filled with moments that loop back on themselves, creating a beautiful chaos that challenges our everyday understanding. This is the fascinating place where infinity meets reality.

When we think about supertasks—tasks that involve an endless number of steps—our minds start to buzz with curiosity. Do such tasks really exist in our universe? The very idea of supertasks raises questions that touch on the heart of existence. In a world governed by physics, where particles and waves dance together in a delicate balance, can we really accept the idea of doing an infinite number of actions within limited boundaries?

Let's explore this idea a bit more. Picture a universe where supertasks can happen seamlessly, much like Thompson's Lamp that we discussed earlier. In the strange world of quantum mechanics, we're used to things not working the way we normally expect. Quantum phenomena often break the rules of cause and effect, where particles can exist in many states at once, a concept known as superposition. If particles can behave as if they're flirting with infinity, can we also consider that supertasks might have a role in our reality?

Think about the on-off paradox we examined before. If we look closely at how particles act at the quantum level, the possibility of being both "on" and "off" at the same time becomes fascinatingly believable. This strange duality challenges our traditional view of reality and nudges us toward an understanding where infinities coexist. In a way, reality might be more complicated than we ever thought. Perhaps the universe is a bit like Thompson's Lamp, flickering between states, each flick representing a different possibility—all existing at once.

But let's not just theorize. In the field of physics, we come across time dilation—a mind-blowing concept that shows how time can stretch or shrink based on speed and gravity. Imagine a scenario with twins: one twin takes off on a space adventure, zooming through the stars at near-light speed, while the other stays on Earth, living a typical life. When the traveling twin comes back, they discover that their sibling has aged a lot more than they have. This reveals something profound; time is not a fixed entity but something that can bend and change depending on circumstances. This idea shows us that the infinite nature of time itself can lead to very real changes in our lives.

Infinity Paradoxes

Now, let's think about the deeper meanings behind these ideas. If time isn't a constant, how can we define existence? What if each moment can be viewed as a supertask—where every tiny slice leads to countless outcomes? What does this mean for our understanding of choice, cause and effect, and the nature of reality? The stories we tell ourselves about our lives, shaped by decisions and consequences, create a rich landscape of endless possibilities.

Let's paint a picture to illustrate this concept further. Imagine the universe as a massive movie set, where every scene can be rearranged infinitely. Every character, every event, and all the dialogue aren't set in stone but can unfold in countless ways. The director—representing the laws of physics—has the power to change the timeline, cut scenes, and create new narratives without disrupting the overall story. In this light, our existence is like a film that can be edited endlessly, where every choice spins a new thread in the fabric of time, and where time itself is layered like a complex narrative.

As we sort through the complexities of infinity and reality, it's easy to feel a bit dizzy. The questions start piling up: What if our choices are just one of countless scripts being played out at the same time? What if our sense of time is merely an illusion created by our limited understanding? These thoughts can be unsettling, even confusing, but they also ignite a sense of wonder about our role in this vast cosmic performance.

Now, let's tie these ideas back to the world of physics and the wonders of quantum mechanics. The idea of entanglement, where particles become connected no matter how far apart they are, adds another layer to how our universe works in ways that challenge our old beliefs. When two particles are entangled, a change in one immediately affects the other, even if they are light-years apart. This phenomenon seems to break the rules of time and space, suggesting that reality itself is woven together in a way we're only starting to understand.

With this in mind, we can return to Thompson's Lamp and its paradox. The lamp flickers between on and off, yet it captures the dual nature of existence. Similarly, our understanding of reality may not depend on simple yes-or-no choices—what is or isn't—but instead exists within a vast range of possibilities. Just as the lamp represents an infinite series of states, our reality might depend on the endless variations of existence that we can't fully grasp.

As we explore these mysteries, it's important to reflect on the implications of what we're discovering. If reality is indeed a sprawling movie set where every choice echoes throughout the universe, it raises an intriguing question: what part do we play in this narrative? Are we just viewers, or do we actively influence the unfolding story of the universe? The dance of infinite possibilities also makes us think about agency. Can we shape the outcomes, or are we simply following a script written by the laws of physics?

This exploration doesn't just challenge our views on existence; it also encourages us to embrace a spirit of curiosity. The universe, in its infinite wisdom, invites us to question reality and seek truths beyond the obvious. It inspires us to consider our role in this grand cosmic story and to approach life with a sense of wonder, curiosity, and purpose.

As we journey through our lives, let's hold onto the essence of this exploration. Each moment, just like the flicker of Thompson's Lamp, holds the potential for endless outcomes. Every choice, every thought, and every action is a switch that shapes our reality. In this dance of light and shadow, we find ourselves intertwined in the very fabric of existence, continually crafting and recrafting our story.

So, the next time you find yourself wondering about the nature of reality or the mysteries of time, take a moment to appreciate the infinite possibilities that lie before you. Embrace the curiosity that comes from questioning established norms and the courage to explore what lies beyond our current understanding. As we reflect on our place in this boundless universe, may we nurture a sense of wonder that inspires us to keep seeking answers to the fascinating, intricate dance between infinity and reality. This journey, filled with uncertainty and excitement, is ultimately what makes being human so richly rewarding.

Chapter 5: Gabriel's Horn – A Finite Volume with Infinite Surface

Creating the Horn

Imagine a trumpet-shaped object that exists more in the world of ideas than in our everyday lives. This fascinating shape is called Gabriel's Horn, and it isn't just a drawing on a page; it presents a puzzle that sparks curiosity and pushes us to think about dimensions, surfaces, and volumes in new ways. The idea that something can have a finite volume while its surface area stretches out infinitely is a mystery that has captured the interest of mathematicians and thinkers for ages.

To really understand Gabriel's Horn, let's start with how it is created. The horn comes from a simple mathematical curve defined by the equation $y = 1/x$. If we think of it in calculus terms, it's the graph of a hyperbola. When we plot this on a coordinate system, the curve gracefully drops toward the x-axis, getting closer and closer but never quite touching it. It hovers on the brink of infinity, creating a sense of wonder and intrigue.

Now, let's use our imagination to picture what happens when we spin this curve around the x-axis. This action forms a three-dimensional shape that looks like a trumpet or horn, earning it the name Gabriel's Horn. But unlike a regular trumpet you might come across in a band, Gabriel's Horn has an astonishing property that seems to break the rules of physics: it has a finite volume but an infinite surface area.

To wrap our heads around how this works, visualize the process of crafting the horn. Think of yourself as an artist standing before a large canvas. Instead of painting a picture, you're guiding an imaginary line that stretches endlessly along the positive x-axis. As you trace the curve of $y = 1/x$, you can imagine the motion of creating the horn. If your brush were infinitely long, it would keep painting the curve endlessly, looping around itself until it shapes the surface of the horn.

Now, let's pause to consider what this image really means. When we calculate the volume of Gabriel's Horn—something we can do by integrating the areas of circular slices along the x-axis—we find that it's a finite amount. The math leads us to a clear number, demonstrating that, surprisingly, Gabriel's Horn can hold a definite amount of liquid, no matter how big that volume appears.

But the real twist comes when we look at the surface area. When we perform the integration to find this, we discover that it approaches infinity. This means that if we could somehow 'unwrap' the horn and lay its surface flat, the area it would cover would be so vast that it's hard to even grasp. The idea that a shape can contain a limited volume while needing an infinite amount of material to cover its surface challenges our understanding of mathematics and reality.

To make this concept clearer, let's think about a familiar item: a gallon of milk. Imagine you have a gallon of milk ready to pour into a vessel shaped like Gabriel's Horn. Since the horn's volume is finite, you can easily fill it up with the milk. However, if you wanted to create a label for the horn's surface, the amount of ink needed to cover it would be infinite—assuming you could paint it without running out. It's at this intersection of the finite and the infinite where our perception of reality starts to stretch.

This intriguing paradox isn't just confined to the abstract world of math; it often pops up in our everyday lives, challenging how we see the world. In a universe governed by physical laws, the idea of infinite surface area is hard to reconcile with what we experience through our senses. Yet, here we have a mathematical concept that thrives in this tension, prompting us to rethink our definitions of 'real.'

The history of Gabriel's Horn adds an interesting layer to this discussion. This mathematical curiosity traces back to the 17th century and the work of mathematicians like Evangelista Torricelli, a student of Galileo. Torricelli was captivated by the properties of solids formed by revolution, setting the stage for future explorations into the nature of infinity. His findings left a lasting impact on mathematics and philosophy, echoing through the ages.

As we explore Gabriel's Horn, it's vital to realize that the implications of this paradox extend beyond pure math. They invite us to ask fundamental questions that challenge our understanding of space, size, and the universe. What does it mean for something to have a finite volume yet an infinite surface area? How can we reconcile this with our usual experiences of the world?

On a broader level, the paradox of Gabriel's Horn raises important questions about the limits of human understanding. As we look into the concept of infinity, we inevitably face the edges of our comprehension. Are there aspects of existence that will always remain just out of reach, much like the elusive curve of $y = 1/x$ as it approaches the x-axis?

As we delve deeper into the wonders of Gabriel's Horn, we'll uncover not only its mathematical foundations but also its philosophical implications and real-world applications. The beauty of this paradox lies not just in its numerical properties but in its ability to inspire awe and curiosity about the endless possibilities that await us.

To truly appreciate Gabriel's Horn, we should also acknowledge the mathematical tools that let us engage with such a complex idea. The language of calculus, with its integrals and limits, serves as a bridge between the abstract and the tangible, helping us measure and manipulate concepts that initially seem beyond our understanding.

Even as we seek to simplify Gabriel's Horn, we find ourselves drawn deeper into the mysteries of infinity, where every answer leads to even more questions. What other mathematical wonders might be waiting for us, resonating with the paradoxical themes we've encountered so far? What secrets lie beneath the surface, just as the infinite area of the horn hides behind its finite volume?

Our journey into the world of Gabriel's Horn is just getting started, and the promise of mathematical discovery invites us to dig deeper. We may discover that the true allure of mathematics doesn't just lie in its answers but in the intriguing questions that arise as we strive to understand the universe in all its infinite complexity. Let's keep our minds open and be ready to embrace the unexpected twists and turns waiting for us on this captivating mathematical adventure.

Calculating the Paradox

Picture this: you pop open a bottle of soda, and the fizzy sound fills the room as bubbles surge toward the surface, eager to break free. Pouring the soda into a glass, you can almost see the bubbles dancing upward—a lively and chaotic performance. If we take this everyday moment and translate it into mathematics, we land at an intriguing intersection where ideas like volume and surface area take on a fresh perspective, much like the fascinating paradox of Gabriel's Horn.

Just like measuring the soda in that bottle, we're going to examine the volume of Gabriel's Horn using a method called integration. While that term might sound a bit daunting, think of integration as a way to fill a container—drop by drop. Each drop is like a tiny slice of volume, and when we gather them all, we create the full picture. This analogy helps us understand the key idea of adding up endless quantities to reach a definite result, which is crucial for understanding Gabriel's Horn.

Mathematically, Gabriel's Horn is described by the function $y = 1/x$, with x stretching from 1 to infinity. Visualize a curve that smoothly slopes down toward the x-axis, getting closer but never actually touching it. If we spin this curve around the x-axis, we create a three-dimensional shape—the horn itself. To figure out the volume of this horn, we'll use the disk method, which involves slicing the horn into infinitely thin circular disks.

The volume V can be found using this integral formula: $$V = \pi \int_{1}^{\infty} (1/x^2) \, dx$$

In this case, $1/x^2$ tells us the area of each circular slice, while π accounts for the shape when it spins around the x-axis.

Now, let's break this down into simpler parts. We want to find the volume of a horn extending from x = 1 to x = ∞. The integral sums up all those incredibly thin disks starting from x = 1 and going on to infinity.

To solve the integral, we need the antiderivative of $1/x^2$, which is -1/x. We can rewrite our volume formula like this: $$V = \pi \left[-\frac{1}{x}\right]_{1}^{\infty}$$

When we evaluate this at our limits, we first substitute the upper limit, which is infinity. As x moves toward infinity, -1/x gets closer to 0. Then we substitute the lower limit, which is 1. So we get: $$V = \pi \left[0 - (-1)\right] = \pi$$

And just like that, the volume of Gabriel's Horn is π, a finite number! This means if we filled the horn with liquid, it would hold exactly π cubic units. Just like the soda in that bottle, we've managed to contain something significant within what seems like an endless shape.

But hold on—what about the surface area? This is where the paradox really starts to unfold. To find the surface area A of Gabriel's Horn, we use a different formula: $$ A = 2\pi \int_{1}^{\infty} (1/x) \sqrt{1 + (dy/dx)^2} \, dx $$

Here, dy/dx is the derivative of our function, which we calculate as: $$ \frac{dy}{dx} = -\frac{1}{x^2} $$

Next, we need to calculate $(1 + (dy/dx)^2)$: $$ 1 + \left(-\frac{1}{x^2}\right)^2 = 1 + \frac{1}{x^4} $$

Now, plugging this back into the surface area formula gives us: $$ A = 2\pi \int_{1}^{\infty} (1/x) \sqrt{1 + \frac{1}{x^4}} \, dx $$

This step can be a bit tricky, but we can make things easier by looking at our integral for large values of x. As x approaches infinity, the term $(1/x)$ gets smaller and smaller, causing the integral to diverge. When we compute this integral, we find that the surface area approaches infinity.

This surprising result can be understood better through our earlier soda metaphor. While the volume of our horn is

finite—like the amount of soda in the bottle—
the surface area stretches infinitely, much like
an endless roll of wrapping paper. Even
though the horn can hold a specific volume,
the material required to cover its entire
surface would have to go on forever, creating
a mind-boggling contradiction that challenges
our intuition.

    To visualize this further, think about a
balloon. When you inflate it, the balloon
expands outward. While the volume of air
inside is limited—just enough to fill the
balloon without popping—the surface keeps
stretching as the balloon grows. Now, imagine
a balloon that could inflate endlessly, with its
surface area increasing but never allowing
more air than what it already contains.
Gabriel's Horn captures this enchanting
contradiction, urging us to rethink our
understanding of space.

    This paradox has captivated
mathematicians, philosophers, and curious
minds throughout history. It encourages us to
explore the edges of mathematical concepts
and prompts us to ponder deeper questions
about infinity, continuity, and our physical
world. How can something be finite in one
way yet infinite in another? What does it
mean to measure and define such entities?

    As we reflect on Gabriel's Horn, we
find ourselves at the fascinating intersection of

mathematics and philosophy, opening our minds to new ways of understanding. The implications of this paradox reach far beyond mere equations; they resonate in the very essence of our existence, challenging us to embrace the mysteries woven throughout the universe.

To further appreciate the wonder of Gabriel's Horn, let's think about some real-world applications of this mathematical curiosity. The idea of having a finite volume and an infinite surface area appears in many natural phenomena, especially in physics and biology. For example, look at certain marine organisms, like coral reefs. These living structures show a finite volume while also having incredibly complex surface areas, offering a habitat for countless marine creatures.

In cosmology, we can consider the universe itself. The observable universe has a finite volume, yet its surface—filled with galaxies and cosmic wonders—could be thought of as infinitely vast. The very nature of space and time bends our understanding of existence, much like Gabriel's Horn tests our grasp of mathematical ideas.

Moreover, the concept of infinite surface area with a finite volume can lead to interesting insights in engineering and design. Imagine creating a container that maximizes the storage capacity while using the least amount of material. Engineers might draw inspiration from the paradox of Gabriel's Horn to come up with designs that use less material for greater capacity, a principle that aligns well with sustainable design practices.

As we dive deeper into this mathematical marvel, it becomes clear that the beauty and complexity of the universe lie in its contradictions. The relationship between Gabriel's Horn and its paradox challenges us to embrace the intricacies of existence. Our imagination often faces tests when we confront ideas that defy tidy categorization. Instead of shying away from these concepts, let's lean in, letting our curiosity guide us into unexplored territories.

Ultimately, the exploration of Gabriel's Horn doesn't just end with a neat answer; it opens up a world of questions. What other mathematical paradoxes might we discover? How do they connect with our understanding of reality? Are there further adventures awaiting just beyond what we can comprehend?

As we ponder these questions, let's keep our minds open and our spirits adventurous. Just as the curves of Gabriel's Horn stretch infinitely, so does our quest for knowledge. In our search for understanding, we find not only the beauty of mathematics but also the deep connections between abstract ideas and the world we experience every day.

Gabriel's Horn isn't just a mathematical concept; it's a lens through which we can examine the world, revealing hidden layers of reality beneath the surface. This paradox invites us to think both critically and creatively, inspiring us to engage with the wonders of the universe with renewed enthusiasm. Whether through the lens of mathematics, philosophy, or natural sciences, Gabriel's Horn acts as a bridge to the infinite, reminding us of the thrill of exploration and the joy of discovery.

Real-World Connections

Picture yourself chatting with an architect as they lay out the plans for a stunning new building that pushes the boundaries of modern design. The architect spreads blueprints across the table, each page bursting with innovative ideas that promise to marry beauty with functionality. While the vision might seem almost magical, it's important to remember that behind every architectural wonder are solid mathematical principles, especially when it comes to understanding volumes and surfaces. One such principle is Gabriel's Horn—a fascinating yet puzzling concept that, when we look at its real-world applications, reveals how deeply mathematics connects with our everyday lives.

At first, the idea of a horn that has a finite volume but an infinite surface area might come off as just a quirky thought experiment, something you'd only find in a textbook. But as we dig a little deeper into this paradox, we see that it plays a role in various fields such as physics, engineering, and even biology. Each of these areas wrestles with the strange ideas of infinity, finite measurements, and the beautiful realities that follow. They remind us that math is not merely a collection of formulas but a vital tool that helps us make sense of our world.

In engineering, designing buildings and materials often hinges on calculating volume and surface area. When engineers set out to create objects that are both lightweight and capable of holding large amounts, they often rely on geometric shapes that echo the principles shown by Gabriel's Horn. Take, for example, a water tank. Engineers must ensure that the tank can hold a specific amount of water (volume) while also making sure it can handle outside pressures, which involves figuring out the surface area.

By applying the lessons from Gabriel's Horn, engineers can enhance their designs. They might opt for curved surfaces that mimic the horn's graceful shape, which can improve how fluids move and reduce the amount of material they need. In this way, Gabriel's Horn transforms from a theoretical idea into a useful guide for solving real-world problems. The concept of finite volume helps engineers confine liquids within structures while also reminding them that they can manipulate surface area to achieve better performance without needing endless amounts of materials.

Another exciting area where the paradox of Gabriel's Horn shines is in fluid dynamics. Imagine water flowing through pipes and the intricate behaviors that arise as this fluid interacts with different surfaces. Engineers design systems that must effectively manage this flow while reducing turbulence and energy loss. Here, the relationship between surface area and volume is crucial. The principles that determine how water behaves against the inside of a pipe can be likened to the infinite surface area of Gabriel's Horn, showing that despite its finite volume, it presents unique challenges and opportunities for smart design.

Think about the water moving through the curves and bends of a pipe. As it flows along, it interacts with the pipe's material—an interaction that can be examined through calculus and physics. Engineers might model this flow to see how changes in surface area affect resistance and speed. Just as Gabriel's Horn has a surface area that approaches infinity, the physical behavior of fluids reveals complexities that engineers must consider. Understanding these interactions helps improve efficiency, reduce energy use, and influences the creation of cutting-edge technologies.

The charm of Gabriel's Horn isn't limited to engineering; it also finds its way into the natural world, particularly in the study of certain biological structures. Take coral reefs, for example. These underwater wonders are filled with intricate designs that inspire awe while showcasing nature's incredible beauty. Coral reefs have a finite volume, yet their surface area is remarkably extensive. Every tiny crevice and nook within a reef provides a home for countless marine creatures. The complex surface of the reef supports a rich diversity of life within a relatively small space, much like Gabriel's Horn, which features finite volume and infinite surface area.

Coral polyps, the tiny animals responsible for building these reefs, display a remarkable ability to create structures that maximize their access to nutrients and sunlight. Walking through a coral reef feels like wandering through an underwater garden, with every turn unveiling new life. The geometry of the reef mirrors the paradox of Gabriel's Horn, illustrating how limited resources can come together to create a vibrant ecosystem. By appreciating the lessons from this mathematical idea, we can better understand the delicate balance of nature, where finite elements come together to form a thriving habitat.

As we consider these real-world connections, it becomes clear that the lessons from Gabriel's Horn go beyond mere academic interest; they challenge us to look at the world in a new light. This paradox prompts us to think carefully about how we measure and understand space, volume, and surface area. Beyond their geometric implications, these ideas encourage us to explore the philosophical questions surrounding infinity and existence itself. How can something be both finite and infinite? What does it mean to measure something that seems immeasurable?

These reflections resonate across various fields, sparking discussions that span mathematics, physics, biology, and philosophy. The work of scientists and mathematicians through the ages shows just how relevant these ideas remain. Think of influential figures like Galileo and Leibniz, who grappled with the challenges of infinity and helped lay the groundwork for calculus. Their work reminds us that the pursuit of knowledge is a journey that invites us to embrace questions, knowing that we don't always have to find immediate answers.

Engaging with paradoxes like Gabriel's Horn enriches our understanding of the world, encouraging us to apply mathematical insights to real-life challenges. Whether we're designing sustainable buildings, optimizing fluid systems, or admiring the beauty of coral reefs, the principles drawn from this paradox connect with our everyday experiences. They deepen our appreciation of nature and inspire creative thinking in engineering and beyond.

As we carry these insights forward, we can look to Gabriel's Horn to spark our curiosity about the endless possibilities around us. This paradox encourages us to keep an open mind, exploring the fascinating links between space, volume, and surface area. Just as the horn extends infinitely, so does our quest for knowledge, propelling us to explore the mysteries that shape our universe.

Our journey doesn't end with just one understanding; it evolves as we make new discoveries. By bridging mathematical theory with real-world applications, we embrace the richness of both realms, igniting a curiosity that encourages us to dig deeper into the endless landscape of ideas.

The connections we form through concepts like Gabriel's Horn remind us that mathematics isn't just about numbers and equations; it's a way to understand the universe in all its complexity and wonder. From the intricate designs of coral reefs to the brilliant creations of engineering, the impact of this mathematical paradox resonates throughout our lives, encouraging us to explore the world with a relentless curiosity and a desire to uncover the beauty that lies within the fabric of life.

Ultimately, the legacy of Gabriel's Horn isn't just in its mathematical elegance; it's also in the meaningful connections it fosters across different fields. It teaches us to embrace the mysteries of infinity and to approach problems with creativity and innovation. As we navigate through the complexities of life, we can find harmony between the finite and the infinite, recognizing that within this paradox lies an invitation to explore, discover, and reflect on the vastness of our world. By embracing this journey, we can uncover the beauty at the crossroads of math and the reality we encounter every day, guiding us toward a deeper understanding of both ourselves and the universe that surrounds us.

Chapter 6: The Ross-Littlewood Paradox – The Vase That Never Fills

Infinite Processes

Picture a vase resting on a table, a beautiful piece of porcelain that catches the light just right. You might find yourself curious about its purpose. Is it meant to hold flowers, serve as decoration, or could it be something much more intriguing—perhaps a doorway to understanding infinity? Now, let's stir things up a bit. Imagine that every time you drop a ball into this vase, another one is taken out at the same time. But here's the twist: both of these actions happen endlessly.

This isn't just a fancy thought experiment; it perfectly captures what's known as the Ross-Littlewood Paradox. As we explore this paradox together, we'll uncover the workings of infinite processes, revealing how they can lead to results that surprise our everyday thinking about math.

To better grasp the paradox, let's first set some ground rules for our vase. We begin with an empty vase, like a blank slate just waiting to be filled. The first action is to drop a ball into it. At the same moment, we're going to remove a ball. If we keep this process going forever, you might ask—what ends up in the vase? Common sense tells us that after adding and removing an infinite number of balls, the vase should have some balls in it, right?

But here's the big surprise: the vase stays empty.

At first, this conclusion might feel a bit strange. How can we add and take away infinitely, and still end up with nothing? To understand this, we need to think about what it means to do something infinitely. In math, infinity isn't just a giant number; it's a tricky idea that goes beyond our usual way of thinking. When we say we're doing something infinitely, we're talking about a process that goes on and on, a series that never stops.

Let's use a simple analogy. Imagine filling a cup with water. As you pour, the cup fills up, and you can see it almost overflowing. Now, think about an infinite version of this: for every drop of water that goes in, a drop also flows out. If we keep pouring and removing drops forever, will the cup ever fill up? No, because for every drop added, there's one being taken away. The result is a cup that never fills, no matter how much water you pour in.

The Ross-Littlewood Paradox asks us to think about this idea of infinite processes. It makes us question what we think we know about accumulating things. We like to see addition as a simple act with clear results, but infinity plays by entirely different rules.

Let's break down what happens in our vase scenario. Each time we add and remove balls, we create an endless series of actions. We drop the first ball in, then we take the first ball out. Next, we add the second ball, and once again, we remove the first ball. This pattern keeps repeating: each new ball is added only to have the first ball taken out again, creating an endless loop.

When we visualize this, it's clear that at no moment can we say the vase holds any balls. Each addition is perfectly canceled by a removal, keeping everything in a steady state of emptiness. Mathematically, this paradox is often explained through limits. As we reach towards infinity, the number of balls in the vase gets closer and closer to zero.

Now, let's explore a variation that highlights how order matters when dealing with infinity. Imagine that instead of adding balls first, we decide to remove all the balls from the vase before adding new ones. In this case, if we take the balls out endlessly before we start adding, the vase ends up empty. But if we add balls first, even with the removal that follows, we'd find ourselves with an infinite number of balls in the vase.

This leads us to an interesting point: the order in which we do things can drastically affect the outcome, especially in the world of infinity. It shows a fundamental truth about math—changing the order can lead to very different results. The Ross-Littlewood Paradox illustrates this principle beautifully, showing how our understanding of addition, subtraction, and infinity can shift dramatically depending on how we approach it.

As we keep exploring the implications of the Ross-Littlewood Paradox, it's crucial to remember that this isn't just a playful math puzzle. It touches on deeper philosophical questions about reality and existence, and the systems we build to make sense of the world around us. The vase that never fills is more than a theoretical idea; it encourages us to rethink how we perceive existence itself.

In our daily lives, we encounter infinite processes in many forms, from the cycle of day turning into night to the seemingly endless flow of time. Infinity surrounds us, yet it's still a tough concept to fully grasp. The Ross-Littlewood Paradox, with its whimsical vase, reminds us that our instincts about reality can sometimes lead us astray.

Let's pause for a moment and think about the importance of these infinite processes. As we journey through life, we often face situations that appear simple on the surface but can unravel into complexities if we look closer. The paradox of the vase might mirror our experiences. We may feel like we're gathering knowledge, forming relationships, or collecting experiences, yet upon deeper inspection, we might discover a balance that keeps us in a state of constant motion rather than true accumulation.

In the larger picture, the idea of the vase that never fills encapsulates a broader narrative about life and existence. It pushes us to ask what it really means to fill our lives with meaning, purpose, and fulfillment. Are we genuinely adding value, or are we stuck in an endless cycle of seeking without truly finding?

By engaging with the Ross-Littlewood Paradox, we not only deepen our grasp of mathematical ideas but also dive into a philosophical examination of infinity, existence, and how we see reality. The vase stands as a strong symbol, urging us to reflect on our actions and the results they bring.

As we navigate the complexities of infinity, the vase serves as a reminder that some processes yield results that surprise us, encouraging us to face life's paradoxes with curiosity and an open heart. The infinite nature of our experiences, much like the operations we perform on the vase, challenges us to rethink how we approach knowledge, existence, and the very fabric of reality.

So, whether we're pondering the workings of math or reflecting on the essence of our lives, the Ross-Littlewood Paradox invites us to engage in a richer understanding of infinity and its deeper meanings. We come away with the insight that infinity is not just a math concept but also a philosophical journey that challenges how we perceive the world, urging us to explore, question, and grow in our understanding of ourselves and the universe we inhabit.

Order Matters in Infinity

Imagine a game of pure imagination, where players take turns with a vase, a collection of balls, and a set of rules that might leave even the best mathematicians scratching their heads. The players are a mix of curious children and wise elders, all excited to explore this world of endless possibilities. In this game, the outcome depends not just on how many balls are added or taken away, but also on the exact order of these actions. It's like a playful dance, showcasing the fascinating quirks of infinity.

Let's picture the scene: there's a vase that starts off empty. Player A adds a ball to the vase, and right after, Player B removes a ball. What happens next? If they keep taking turns—adding and removing balls endlessly—players might think they're building up a pile of balls in the vase. But surprisingly, the vase stays empty! This logic feels strange and makes you question how infinity can work this way.

Now, let's change the rules a bit. Imagine Player A decides to take all the balls out of the vase before adding any. If they remove every single ball and then start adding, wouldn't the vase still be empty? The answer is a definite yes! Even in situations that seem simple, the order of actions can lead to unexpected results. This shows us something remarkable: in the world of infinity, order really matters.

As we dive deeper into this playful exploration, we can mix things up with different sequences of actions. Picture a scenario where the players add two balls to the vase before taking one away. The result shifts once again. If we repeat this process forever, the vase ends up containing an infinite number of balls! Just a tiny change in actions can lead to dramatically different results.

So why does all this matter? Infinity has a unique way of challenging our everyday thinking. In our daily lives, we usually think in a straight line—one action leads to a clear result. But with infinity, that simple approach falls apart. The rules we rely on become blurry, and that's where the magic, or perhaps the chaos, of infinity comes to life.

To illustrate this further, let's imagine a lineup of players where each one represents a different ordinal number, a way to describe sizes of infinity. In this setup, the players are numbered 1, 2, 3, and so on, stretching out to infinity. Player 1 adds a ball, Player 2 takes it away, then Player 3 adds another. Each player's action affects the overall outcome.

But what if we mix things up and let Player 2 go first, removing a ball before Player 1 adds one? Just like that, the contents of the vase change in an instant! The key point here is recognizing that the order of the players matters a lot. What seems like a small detail can reveal the complex nature of infinite sets and how they interact.

This playful game illustrates a vital principle of infinity: the order of operations can create vastly different realities. The Ross-Littlewood Paradox shows us how our intuitive understanding of addition and subtraction can falter when faced with infinity. These findings challenge our long-held beliefs and push us to rethink our understanding of reality.

Now, let's invite a new character into our imaginative game: a friendly, mischievous cat named Whiskers. Whiskers loves to mess with the balls in the vase, adding a layer of chaos to our game. Sometimes, Whiskers drops a ball in, only to swat it out a moment later. This playful distraction leads to unpredictable results. As the players keep adding and removing balls, Whiskers' antics change the dynamics of their game.

As Whiskers adds and removes, the players find themselves in a tricky situation. How do they keep track of everything? Whiskers is like an infinite wildcard, reminding us that chaos is a natural part of infinity. This delightful distraction illustrates an important idea: in the world of infinite processes, unexpected changes can lead to even more surprising outcomes.

Now let's imagine another twist: instead of taking away balls, the players add them at an increasing rate. Player A adds one ball, Player B adds two, Player C adds three, and so on. As this continues indefinitely, the vase fills up faster than anyone can manage. It doesn't just hold an infinite number of balls; it starts overflowing, teetering on the edge of chaos. This scenario shows how scaling up actions can affect the final outcome, underscoring the significance of order and rate in infinite operations.

The game keeps evolving, with players trying different sequences of actions, where seemingly innocent changes lead to a wide array of outcomes. Each twist and turn deepens our understanding of infinity and highlights just how much order matters.

Looking beyond our playful game, we should think about the bigger implications of these ideas. Consider how order plays a role in our everyday lives. We often think of events as linear, where one action leads directly to another. However, as the vase game has demonstrated, altering the order can create entirely different realities.

Take organizing your schedule, for instance. You might believe that prioritizing tasks based on how urgent they are is the best way to manage time. But if you're always juggling tasks, the most important ones might end up getting pushed aside. Similarly, in complicated systems, the order of operations can determine whether a project flourishes or fails.

In the grand scheme of life, these principles extend to our relationships, personal growth, and the search for knowledge. Often, we chase experiences, trying to fill our lives with meaning. But if we go about things without thinking about the order—without reflecting on our choices—we might find ourselves stuck in routines that lead to emptiness rather than fulfillment.

As we navigate the complexities of life, the vase stands as a symbol of our efforts. It encourages us to think carefully about our choices and how they shape our reality. If we learn from the Ross-Littlewood Paradox and our imaginative game, we must acknowledge the powerful influence of order in our choices and lives.

In this journey through infinity, we've encountered the playful nature of sequences and operations. But perhaps the most important lesson is to embrace complexity. The interaction of actions highlights that understanding infinity isn't just a math puzzle; it's a chance for personal growth and a deeper grasp of our world.

As we keep exploring these ideas, let's allow our imaginations to soar. Picture a world where the order of our actions opens doors to new realities. Each decision creates a web of possibilities, a colorful array of outcomes that challenge our usual way of thinking.

By engaging with the playful dynamics of our imagined vase game, we not only grasp the intricacies of infinity but also reflect on the importance of our daily choices. Each ball added or removed is like our actions in life, emphasizing the need for careful thought and mindfulness.

In the end, the lesson we take from this delightful exploration is more than just a fun math trick. It's a reminder of how our understanding of order shapes our existence, encouraging us to approach life with wonder and curiosity. The infinite possibilities dance before us, inviting us to engage with the complexities of our reality and to question our experiences with open hearts and eager minds.

Challenging Intuition

When you start to think about infinity, it doesn't take long before your mind hits a wall where intuition gets shaky and the clear lines of reasoning start to blur. The Ross-Littlewood Paradox illustrates this perfectly, pushing us to rethink our instincts about math. It sets up a situation where two different sequences of actions can lead to contradictory truths regarding an infinite collection. How can something be true in one context but false in another? This puzzling inconsistency shows just how complicated it can be to wrap our heads around infinity, raising deep questions about the very nature of mathematical reality.

At its heart, the paradox brings our intuitive grasp of infinity into question. We often think of numbers as solid and reliable, but infinity is a slippery idea that slips through our fingers and makes us rethink how we understand truth in math. It's a philosophical puzzle that nudges us to challenge our assumptions and reconsider what we believe about existence itself. As we unpack the implications of the Ross-Littlewood Paradox, we can't help but wonder: What does it tell us about the nature of mathematical truth and reality?

Let's take a moment to think about the early thinkers who boldly tackled the mystery of infinity. Georg Cantor, the pioneer of set theory, introduced the concept of different sizes of infinity, distinguishing between countable and uncountable infinities. His groundbreaking ideas transformed not just mathematics, but also sparked debates that still echo today. Cantor's work opened up a whole new way of looking at infinity, revealing it as a rich landscape filled with complexities instead of a single, simple idea.

Then there's Zeno of Elea, whose paradoxes continue to provoke thought about infinity. His famous dichotomy paradox claims that motion is impossible because you must first reach the halfway point, which questions our basic notions of space and time. Zeno's theories remind us that when we view things through the lens of infinity, our intuitions about movement and progress might break down. The ideas of these thinkers blend together with the Ross-Littlewood Paradox, creating a fascinating philosophical backdrop that encourages us to dig deeper.

As we explore these ideas, take a moment to reflect on your own encounters with infinity. Have you ever looked up at the stars on a clear night, marveling at the seemingly endless sky? In those moments, the vastness can fill you with a sense of awe and even insignificance. Or maybe you've become absorbed in a mathematical concept that spirals out into infinite possibilities, revealing all sorts of complexities that challenge your understanding. These experiences remind us that infinity is not just an abstract idea in math; it seeps into our everyday lives and invites us to rethink how we perceive reality.

Think about time, for example. We often see it as a straight line: past, present, and future, all neatly arranged. But when we start to consider the infinite nature of time, that neatness starts to fade. What if time isn't a straight path but instead a vast web of possibilities? This idea shakes up our intuition about how we live and experience life, prompting us to question the very essence of our reality. The Ross-Littlewood Paradox pushes us to embrace this uncertainty, to challenge what we believe and explore the many ways in which infinity affects our understanding of the world.

As we think about the significance of the paradox, we realize it raises big questions about the connection between math and truth, and even the nature of reality itself. If a mathematical statement can be true in one situation and false in another, how do we define truth in an infinite setting? Is truth something absolute, or does it change depending on the circumstances?

This isn't just an academic question; it resonates deeply in our everyday lives. Think about how truth can be fluid in our relationships, where what one person sees as betrayal, another might see as a misunderstanding. The variety of truths reflects the complicated mix of emotions, experiences, and interpretations that shape our interactions. Is it possible that the world of mathematics is also influenced by these complexities?

As we navigate this intricate landscape, it's important to appreciate the fascinating yet confusing nature of infinity and its paradoxes. The Ross-Littlewood Paradox, with its playful challenges to our thinking, encourages us to dive deeper into our understanding. By recognizing the limits of our intuitive perceptions, we can open ourselves up to new insights and perspectives that may have previously escaped us.

In our exploration of this paradox, let's also think about how paradoxes themselves have driven the evolution of human thought. Throughout history, paradoxes have sparked philosophical and scientific breakthroughs, pushing us beyond conventional thinking. The Ross-Littlewood Paradox is no different; it compels us to look beneath the surface and wrestle with the complexities that come with infinity. Perhaps it's through these confrontations with paradox that we can expand our minds into new territories.

As we ponder the implications of infinity, we're reminded of the value of curiosity and open-mindedness. The journey of learning is rarely a straight shot; it winds through a maze of questions and uncertainties. Engaging with the Ross-Littlewood Paradox invites us to embrace this complexity, reminding us that the process of understanding is just as important as the conclusions we reach. Recognizing the paradoxes of infinity can inspire us to approach our daily lives with wonder and curiosity, pushing us to challenge our intuitions and question our assumptions.

Reflecting on the insights from the Ross-Littlewood Paradox leads us to consider the broader implications for our understanding of mathematics and the universe. Infinity, with all its baffling intricacies, calls us to explore the unknown and confront the limits of our knowledge. It reminds us that our grasp of reality is not static but rather a constantly evolving mix of knowledge, perception, and inquiry.

As we wrap up our thoughts on the Ross-Littlewood Paradox, we find ourselves at a turning point. This paradox challenges us to rethink our ideas about infinity and what it means for truth and reality. It encourages us to question our intuitions, embrace the complexities of the infinite, and engage in a deeper exploration of existence. By nurturing a sense of curiosity and open-mindedness, we can push beyond our limitations and embark on a discovery journey that reshapes our understanding of the world and ourselves.

Avery Pascal

Chapter 7: The Banach-Tarski Paradox – Doubling a Ball Without Breaking Rules

The Paradox Unveiled

Imagine a perfectly round ball, smooth to the touch and as round as a dream. It sits quietly on your coffee table, just another ordinary object in your home. But what if we added a twist of mathematical intrigue? What if I told you that you could take this ball, cut it into pieces, and then somehow put those same pieces back together—without losing any material—into not one, but two identical balls? Yes, you read that right! This isn't just a magic trick; it's a fascinating mathematical riddle known as the Banach-Tarski Paradox.

At first, this idea might feel like a challenge to everything we know about the physical world. How can something exist in two places at once? How is it possible to create a duplicate of a ball simply by slicing it apart? For many, the Banach-Tarski Paradox seems utterly absurd. But if we take a closer look, we can begin to appreciate the beauty of its underlying concepts and the mind-blowing ideas it raises.

At the core of the Banach-Tarski Paradox is the strange world of abstract mathematics. Imagine you take your ball and divide it into a finite number of pieces. But here's the twist—the pieces aren't your typical, neatly shaped slices; they're infinitely complex fragments created by specific mathematical rules. It's hard to even picture these pieces because they challenge everything we think we know about shapes and volumes. In regular geometry, this idea would seem nonsensical. Yet here we are, dancing on the edge of reality itself.

To make this a bit clearer, think about a cake. When you cut it into pieces to share with your friends, each slice is clearly defined; there are no odd shapes or overlaps. Now, imagine you have a magical knife that lets you create pieces that are so unusual they can be reassembled into another cake. This new cake doesn't just appear out of thin air; it seems to break every rule you've learned about volume and space. The Banach-Tarski Paradox is like this whimsical tale from a bakery, where the impossible becomes possible, and the rules we understand start to bend under the weight of mathematical logic.

Now, let's think of your favorite jelly bean. If you were to cut it into pieces, you'd never be able to gather those pieces again to create two identical jelly beans. But in the abstract world of mathematics—where the Axiom of Choice reigns supreme—that sort of thing becomes doable. The Axiom of Choice is like a magical key that allows us to explore the infinite and the unimaginable, making it possible to create these strange pieces that don't exist in our physical world.

The ripple effects of the Banach-Tarski Paradox reach deep into math and philosophy. It pushes us to rethink our understanding of infinity and the concept of volume. Are we really limited by the rules of the physical world when mathematics allows such wild conclusions? This paradox doesn't just shake up our understanding of geometry; it also prompts us to reconsider the very essence of set theory and what it means to exist.

As we dive deeper into this paradox, it's important to recognize the discomfort that can come with it. Behind every tricky math puzzle is the chance for clarity and insight if we're open to facing it directly. The Banach-Tarski Paradox invites us to look beyond our usual perceptions and embrace a universe where logic and intuition don't always go hand in hand. It's a call to explore the endless possibilities that lie just beyond the surface of our daily lives.

What's truly captivating about this paradox is how it exposes the limits of our understanding. Imagine a world where mathematical ideas operate under rules that seem to contradict our everyday reality, where you can take one object and create duplicates from nothing. This concept can be unsettling, but it's also thrilling. Like a gripping story that makes you question the nature of existence, the Banach-Tarski Paradox encourages us to rethink what we believe about the universe.

The implications of this paradox go far beyond mathematics. They touch on philosophical questions about reality, existence, and our grasp of the infinite. As you ponder these ideas, remember that many mathematicians and philosophers over time have wrestled with the meanings behind this paradox. They've tried to balance the alluring beauty of math with the concrete reality we live in. This tug-of-war between these two worlds invites a richer conversation about knowledge itself.

As we unveil the secrets of the Banach-Tarski Paradox, let's take a moment to appreciate the art found in mathematical abstraction. In this realm, we can venture into ideas that stray from our everyday experiences. The idea of infinity becomes more than just a big number; it turns into a playground for our imagination. You'll find that the paradox is less about the literal act of creating two balls from one and more about exploring concepts that challenge our perceptions.

Infinity Paradoxes

The beauty of mathematics lies in its power to inspire wonder. The Banach-Tarski Paradox stands as a shining example of this, inviting those who are curious enough to engage with its implications. It nudges us to think about existence, the limits of our understanding, and the potential for infinite possibilities. So, the next time you glance at a simple ball on your table, remember that beneath its ordinary exterior lies a world of extraordinary ideas waiting to be discovered.

Welcoming the paradox helps to shed light on the deeper layers of mathematical thought. It sparks a curiosity that is essential for understanding the world around us. While we might not actually slice our physical objects to create duplicates, the mental exercise is crucial for growth and exploration. The Banach-Tarski Paradox reminds us that the beauty of mathematics often comes from the unexpected, and it's that unpredictability that drives our thirst for knowledge and understanding.

As we wrap up our journey into this paradox, let's keep in mind that mathematics isn't just about numbers and equations. It's a rich dance of ideas that challenges us to question and probe the very fabric of reality. The Banach-Tarski Paradox serves as a compelling reminder that the boundaries of math extend way beyond our current understanding, encouraging us to step into a realm where the impossible can become possible. So, as we venture further into the complexities of infinity, let us carry with us the spirit of curiosity that the Banach-Tarski Paradox inspires, ready to embrace the endless wonders that lie ahead.

Axiom of Choice

Picture yourself in a cluttered room, brimming with drawers, each packed to the brim with socks in every shade, design, and size you could think of. You need to grab one sock from each drawer, but here's the twist: you can't peek inside, and there's no clear method for choosing which sock to take. You might feel inclined to dig around blindly, but the Axiom of Choice tells us that even without a visible guideline, it's possible to claim that you can select one sock from each drawer. This straightforward idea opens up an incredible world in mathematics, where such abstract concepts can lead to surprising conclusions.

The Axiom of Choice stands as a key element of modern set theory, acting as an important tool for mathematicians trying to solve problems involving infinite collections. It asserts that when given a group of non-empty sets, we can choose an item from each set, even if there's no specific rule to guide that selection. This means that even in cases where the sets lack a defined structure, the axiom lets us confidently state that a selection function exists. Think of it as a magical tool that lets us pull elements from the vast universe of mathematical possibilities.

To understand the might of the Axiom of Choice, let's look at some well-known mathematical results that spring from its acceptance. One such result is Zorn's Lemma. This principle claims that if every chain (a totally ordered subset) within a partially ordered set has an upper bound, then the entire set contains at least one maximal element. Imagine a hierarchy of sock drawers, each filled with endless pairs of socks, and Zorn's Lemma assures us that amidst this chaotic setup, there's a top sock drawer that holds the ultimate pair—perhaps the fanciest, most stylish socks you can dream up. Zorn's Lemma finds its application in various areas of mathematics, from algebra to topology, revealing the fascinating connections that link seemingly unrelated ideas.

Another important result that relies on the Axiom of Choice is Tychonoff's Theorem. This theorem states that the product of any collection of compact topological spaces is compact. This idea is crucial in many fields of analysis and topology. When we talk about compactness, we dive into a part of mathematics where spaces can behave in surprising ways compared to our everyday experiences. Imagine trying to fit an infinite number of tiny sock drawers into a larger drawer. Tychonoff's Theorem ensures that under the right conditions, you can still manage to fit them all snugly together, despite the infinite nature of your collection.

However, embracing the Axiom of Choice isn't without its controversies. Since the early 20th century, mathematicians have been debating its implications, leading to a kind of divide within the mathematical community. Some view it as a necessary tool for their work, while others push back, arguing that it leads to strange situations that clash with our common sense. Much of the debate revolves around "constructivism," which suggests that mathematical objects should be built explicitly instead of relying on abstract methods offered by the Axiom of Choice. This split opens up rich discussions about what mathematics really is—just a set of useful tools, or something that reflects deeper truths about existence?

Accepting the Axiom of Choice leads us to the astonishing Banach-Tarski Paradox, where the seemingly impossible becomes a reality. This paradox shows how we can take a solid ball, cut it into a finite number of pieces, and then reassemble those pieces into two identical balls. It's one thing to choose a sock from a drawer, but quite another to manipulate the very essence of reality! The Banach-Tarski Paradox prompts us to rethink our notions of volume, space, and even infinity itself. It challenges our views and encourages us to entertain the incredible possibilities that arise from abstract mathematical thinking.

Let's take a moment to appreciate the depth of these ideas. While picking a sock might seem simple, the implications of that choice echo throughout the vast world of mathematics. The Axiom of Choice allows us to explore areas where our intuition might not lead us, blurring the lines of physical reality in the face of abstract reasoning. It's like looking through a kaleidoscope, where you can see a shifting landscape of shapes and colors that defy our everyday understanding.

As we unpack the role of the Axiom of Choice in mathematics, let's consider the idea of a well-ordered set. A well-ordered set is one where every non-empty subset has a least element. The Axiom of Choice guarantees that we can create a well-ordering for any set, even those infinite ones that can leave our heads spinning. This means we can pick a "first" sock from a collection, no matter how chaotic they are arranged. It's a powerful claim that provides a foundational structure to the vast universe of mathematical objects.

But what happens when we push this idea of selection to its limits? What if we apply the Axiom of Choice to an uncountable collection of sets? This is where things start to get particularly strange. The implications go beyond the notion of socks and drawers, diving deep into the sea of infinity. In some instances, this leads to results that challenge our everyday understanding—like the existence of sets that can't be explicitly constructed using standard mathematical methods.

Take the idea of a "non-measurable set," for example. By applying the Axiom of Choice, you can create a set of points that cannot be given a volume in the usual way. It exists in a mathematical sense, but it defies the principles that guide our understanding of measurement. It's a paradoxical entity that only exists because we've chosen to accept the Axiom of Choice, allowing us to explore areas where familiar rules of geometry don't quite fit.

As we continue to explore the implications of the Axiom of Choice, we uncover a landscape that is both wondrous and confusing. It calls mathematicians to examine what exists and the nature of reality itself. Can we truly claim to understand something that exists yet can't be clearly described? The paradoxes that arise from adopting the Axiom of Choice challenge us and invite us to think beyond the limits of our intuition.

While the Axiom of Choice leads to incredible results—like the Banach-Tarski Paradox—it also highlights the delicate balance between abstract mathematics and the tangible world we live in. This paradox serves as a reminder that while mathematical reasoning can unveil astonishing truths, those truths may not hold in our physical universe. The ideas stemming from abstract thought are exciting, yet they also prompt us to question how reliable our intuitions really are.

Consider this: in a universe governed by physical laws, does the Axiom of Choice hold up under scrutiny? If we were to cut our beloved solid ball into pieces, could we actually put those pieces back together into duplicates? The answer reveals the gap between theoretical concepts and practical reality. The Banach-Tarski Paradox, supported by the Axiom of Choice, firmly resides in the realm of theoretical mathematics, where the rules differ from those that govern our everyday lives.

As we navigate this fascinating terrain, it's crucial to remember that the Axiom of Choice isn't just an amusing idea for mathematicians to mull over during their coffee breaks. It's a foundational principle that shapes the core of set theory and has wide-ranging implications across various fields. The discussions surrounding it challenge us to confront some of the most profound questions about existence, infinity, and reality itself.

Delving into the ideas tied to the Axiom of Choice deepens our understanding of mathematics and awakens a sense of wonder about the infinite complexities hidden beneath the surface of our daily experiences. Just like selecting a sock without knowing its color or design, we too can embrace the uncertainty of mathematical abstraction, allowing it to guide us through the mysteries of infinity.

The Axiom of Choice encourages us to dance with the strange and beautiful, a joyful pirouette through the unexpected realms of set theory and mathematical logic. It invites us to expand our minds beyond the limits of what we usually perceive, entertaining the idea that there is a world where the impossible can become possible. As we examine the Banach-Tarski Paradox and its implications, we find ourselves standing at the crossroads of mathematics and philosophy, where boundaries blur, and the infinite calls to us. Accepting the Axiom of Choice opens doors to new understandings, lighting the way to the wondrous complexities of the universe we inhabit.

Implications for Reality

Imagine taking a solid ball, slicing it into a few pieces, and magically piecing those parts back together to create two identical balls. What does this mind-boggling idea tell us about our understanding of the universe? This curious scenario, introduced by the Banach-Tarski Paradox, makes us rethink how we view reality. The idea that we can multiply physical objects just through math isn't just mind-blowing; it challenges how we see the world around us. But while this paradox shows off the creativity of mathematics, it also raises big questions about how math relates to the physical world.

Let's break it down a bit. The Banach-Tarski Paradox works within the realm of pure mathematics, where the rules are a bit looser than those in the physical realm. This paradox relies on something called the Axiom of Choice, which lets us pick elements from infinite sets in a way that isn't too strict. However, when we switch gears from abstract numbers to the real world, we run into the limits set by the laws of physics. You can't just take a rubber ball, chop it up using some math trick, and expect to put together two identical solid balls. The reality we live in is shaped by the properties of matter and energy, like atomic structure and conservation laws, which just don't budge for the sake of math.

Understanding the difference between the world of math and the world we live in is super important. Mathematicians work in a space defined by ideas. They play around with concepts that might seem odd, using theoretical tools that let them explore new frontiers. On the other hand, physicists have to stick to the rules of the universe we inhabit. While mathematicians can toy with infinity and paradoxes, physicists deal with real, measurable things. In a way, the fields of math and physics are like two sides of the same coin—they inform each other while also highlighting their differences.

Take the concept of infinity, for example—a key player in the Banach-Tarski Paradox. Infinity isn't just a number; it symbolizes something limitless, which can be hard to fully grasp. In mathematics, we can break down infinity into different sizes, like countable and uncountable infinity. The paradox acts as a fascinating mirror, showing us how little we might understand about such ideas. If we can manipulate infinity mathematically, what does that mean for our grasp of reality? It challenges us to accept the complexity and uncertainty that lie beyond our immediate understanding, shaking up the very foundations of what we think is possible.

Philosophically, the Banach-Tarski Paradox gets us to reconsider what existence even means. As we dig deeper into infinity and mathematical ideas, we might find ourselves pondering some big questions about life. If we can create two balls from one using math, what does that say about the uniqueness of objects in our world? Can we truly understand a physical object's essence if math allows for such wild transformations? These questions are more than just theoretical; they touch on our understanding of identity, existence, and reality itself.

It's worth noting that the implications of the Banach-Tarski Paradox go beyond mere academic interest. They push us to think about the limits of human understanding. How far can we stretch our comprehension before we're left lost in a sea of paradoxes? This paradox encourages us to embrace uncertainty and accept that some parts of our universe might always remain a mystery. Wrestling with these abstract ideas isn't a waste of time; it's an invitation to broaden our intellectual horizons and confront the wonders that lie just beyond our daily experiences.

Engaging with the Banach-Tarski Paradox also helps us see how it highlights the strengths and weaknesses of human reasoning. Math lets us find elegant solutions to tricky problems, but we need to be careful about applying those solutions to the physical world. The paradox shows the gap between the abstract world of math and the real, practical world we navigate every day. Just because we can imagine an idea doesn't mean it holds true in the universe governed by physical laws. This distinction is crucial for scientists and philosophers alike, shaping how we try to understand our existence.

This discussion also nudges us toward a deeper understanding of the tools we use in mathematics. Think about how we can divide a line segment infinitely into tinier pieces. In theory, that works great. But when we apply it to real matter, we hit challenges. The idea of atoms—finite pieces of matter—limits how far we can divide things down. No matter how much we want to stretch our ideas, we're grounded in the realities of the universe.

Furthermore, the Banach-Tarski Paradox shines a light on the connection between mathematics and the natural sciences. Physics uses math to explain the universe, but it also sets boundaries that pure math doesn't have. This relationship is like a dance, a lively interaction where mathematicians create tools that help physicists, but those tools must still fit with the basic rules of reality. The paradox serves as a reminder that while math is incredibly powerful, it isn't a foolproof way to understand the universe.

These ideas lead us to think about what reality really is. If mathematical ideas can exist in a realm that allows for such strange scenarios, what does that mean for how we see the universe? Can we view reality as a fixed thing, or is it just a collection of perceptions shaped by our limited understanding? The interplay between physics and math invites us to think about our relationship with the unknown, encouraging us to embrace the confusion and wonder that often come with exploration.

As we navigate this intriguing landscape, we should also reflect on how the Banach-Tarski Paradox influences scientific inquiry. It poses a riddle: can we ever fully understand the universe if its fundamental concepts can lead to such bizarre conclusions? This question resonates deeply, reminding us that imagination plays a crucial role in scientific advancement. Just as mathematicians often push the boundaries of conventional thought, scientists too must allow themselves to dream, considering possibilities that might seem far-fetched at first.

In this journey through the implications for reality, we find ourselves standing at an exciting crossroads where curiosity leads the way. The Banach-Tarski Paradox may appear nonsensical, yet it opens up a world filled with potential insights. It challenges our perceptions and urges us to confront our limitations as we strive to understand the universe. Embracing these complexities can guide us to deeper questions that call us to explore the intricacies of existence.

As we wrap up this chapter on the Banach-Tarski Paradox, let's take a moment to celebrate the path we've taken. Our exploration has uncovered not only the mathematical brilliance behind this paradox but also the thought-provoking implications that swirl around it. We've pushed the limits of our understanding, examined infinity, and reflected on our connection with the universe.

Inviting curiosity and exploration is key as we look into the endless possibilities presented by mathematical abstraction. The Banach-Tarski Paradox reminds us that our adventure into the unknown is just getting started. Even if it feels illogical at first, it calls us to embrace the complexities of math and philosophy, encouraging us to seek knowledge that goes beyond traditional thinking. The vast landscape of infinity lies ahead, filled with unanswered questions and mysteries waiting to be discovered, drawing us toward our next adventure in the world of paradoxes.

Chapter 8: The Infinite Monkey Theorem – Typing Shakespeare by Chance

Probability Meets Infinity

Picture a monkey—yes, a monkey—sitting in front of a typewriter, energetically pecking away at the keys. You might chuckle at the silliness of the image, but what if I told you that this quirky scenario is tied to one of the most fascinating concepts in math? Welcome to the Infinite Monkey Theorem, a thought experiment that not only sparks the imagination but also leads us to explore deeper topics like probability, randomness, and the big ideas of infinity.

To get a grip on this theorem, we first need to understand what probability really means. In simple terms, probability is all about measuring uncertainty. It helps us figure out how likely it is for something to happen, with a range from 0 (impossible events) to 1 (certain events). Think about flipping a coin: there are two possible outcomes—heads or tails. The chance of getting heads is 0.5, and the chance of getting tails is also 0.5. This easy example shows how probability gives us a way to make sense of randomness in our daily lives.

Now, let's break down the difference between independent and dependent events, two important ideas that will be crucial as we dive deeper into our monkey's typing adventure. Independent events are those whose outcomes don't affect one another. For instance, if you roll a die, the result (let's say a five) doesn't change what happens when you roll again. Each roll is its own little world of possibilities. On the other hand, dependent events are when the outcome of one event changes the outcome of another. If you're drawing a card from a deck, the chances of pulling a king shift based on whether you put the card back in the deck after drawing it.

With this foundation in place, let's return to our monkey. The Infinite Monkey Theorem playfully suggests that if you set a monkey in front of a typewriter and give it infinite time—yes, infinite time—it will eventually type out the complete works of Shakespeare. It's a strange thought, but it's backed by solid math. Given enough time, every possible arrangement of letters will be typed out. The chance of the monkey randomly producing "To be, or not to be: that is the question" becomes a real possibility. As long as the monkey keeps hitting those keys, the chance of it writing Shakespeare isn't just a silly daydream; it's a genuine mathematical possibility.

Infinity Paradoxes

Let's unpack this a bit more. The complete works of Shakespeare add up to about 884,000 words. The number of ways to arrange those letters is mind-boggling. When you think about the monkey randomly hitting the right keys over an infinite timeline, you'll see that the odds start to lean toward the seemingly impossible. The math shows us something amazing: with infinite attempts, every outcome has a shot at happening.

To put this in perspective, think about the lottery. The odds of winning the Powerball are around 1 in 292 million. If you buy a ticket every week for a year, your chances of winning only improve a little bit; you'd still be facing an almost ridiculous chance of winning that jackpot. But if you could buy tickets forever—imagine, for all time—your chances of winning would eventually get closer to 1. In the realm of infinite possibilities, persistence can lead to the most surprising outcomes.

You might be curious about how this all connects to our everyday lives. Take sports statistics, for example, where probabilities play a huge role. When a basketball player steps up to the free-throw line, their chances of making the shot are often based on their past performance. If a player has a free-throw percentage of 75%, that means over a large number of shots, they're likely to score three out of four times. If we think about infinity in this context, it shows us that over countless games, the player's percentage will stabilize.

This idea of stability in probabilities, especially when considering infinite trials, offers a fascinating way to look at randomness. It helps transform our understanding of events into something more manageable and even predictable, even though they can feel chaotic.

Visual aids can really help make complex ideas clearer, so imagine a bell curve showing a range of random events, illustrating odds and probabilities in an easy-to-understand way. Right in the center are the most common outcomes, while the edges show the rare occurrences. As we stretch the timeline toward infinity, the overall distribution remains the same, but it opens up an endless world of possibilities.

One well-known paradox in the world of probability that often comes up in talks about infinity is the Birthday Paradox. This interesting phenomenon shows how just a small group of people can lead to a surprisingly high chance of shared birthdays. In a room of only 23 people, the odds of two individuals sharing a birthday go over 50%. It seems odd; how can the chances be so high with so few people? This paradox is another example of how our gut feelings can sometimes mislead us when it comes to randomness and probability.

As we start piecing these ideas together, it becomes clear that infinity changes how we see things in profound ways. The concepts behind the Infinite Monkey Theorem don't just stop with the amusing image of a monkey typing Shakespeare. They push us to rethink how we view everything, from unexpected meetings to the very essence of reality.

Looking through the lens of infinity in our daily lives can feel both empowering and a bit overwhelming. But one thing stands out: the unlikely can become possible given enough time. The image of the monkey at the typewriter reminds us that even in the randomness of life, the potential for amazing results is out there. As we move through life, it might seem like the odds are stacked against us, but with the right mindset, we can learn to appreciate the delightful chaos that surrounds us.

As we delve even deeper into the Infinite Monkey Theorem, we'll continue to explore how these ideas show up in real-life situations, looking at how randomness and probability influence our understanding of the universe. The journey through the realm of infinity has the potential to challenge our views, inviting us to embrace the unpredictable and enjoy the possibilities ahead. After all, in a reality where a monkey might just type Shakespeare if given enough time, who knows what other surprises infinity might hold?

Randomness in Action

When we think about randomness, we often picture dice rolling across a table, coins flipping through the air, or maybe even leaves swirling chaotically in a windstorm. We live in a world full of surprises, where events can unfold in ways that catch us off guard. But behind this apparent chaos lies something truly intriguing—a structure and pattern just waiting to be discovered. The flow of randomness isn't just a haphazard shuffle; it's a complex performance that unfolds over time and from different viewpoints.

Let's take a moment to think about what randomness really means. At its heart, randomness is all about the lack of a pattern or predictability. A random event doesn't follow any visible order; its outcomes are independent of what has happened before and can't be predicted with certainty. For example, when you roll a fair six-sided die, the result is completely random, with each face having an equal chance of landing face up. However, randomness doesn't just float around by itself. It interacts with structure, which leads to some fascinating ideas.

In math, we often talk about two types of randomness: true randomness and pseudorandomness. True randomness is the kind we see in nature. Think about how the wind can suddenly change direction or how a raindrop might appear out of nowhere on a sunny day. True randomness is wild and unpredictable; it can't be duplicated or forecasted with absolute accuracy. On the flip side, pseudorandomness is created by algorithms or computer programs. It gives the illusion of randomness, yet it follows specific rules and initial conditions.

A great example of true randomness is found in nature, like the fractals that show up in different natural forms—from the way trees branch out to the unique shapes of clouds. These fractals are self-similar at various scales, showing us that intricate structures can come from simple, random processes. This aspect of randomness can create incredible order, reminding us of a key insight recognized by both mathematicians and scientists: chaos can yield beauty.

The Fibonacci sequence is another fascinating example of nature's knack for forming ordered patterns from randomness. In this sequence, each number is the sum of the two before it: 0, 1, 1, 2, 3, 5, 8, 13, and so on. This straightforward rule creates a sequence that appears all over the natural world, like the way leaves are arranged around a stem or the layout of seeds in a sunflower. It's remarkable how this orderly progression reacts to environmental factors, showing that randomness can lead to predictable results over time.

As we examine how randomness and order play off each other, let's look at chaotic systems, which showcase how unpredictable events can create recognizable patterns. Chaotic systems are very sensitive to initial conditions, meaning that even a tiny change can lead to dramatically different outcomes. This idea is often illustrated by the butterfly effect—the concept that something small, like a butterfly flapping its wings in Brazil, might eventually trigger a tornado in Texas.

Weather patterns are a striking example of chaotic systems. Meteorologists have the tough job of predicting the weather, using complex mathematical models that consider a ton of variables. Because the atmosphere is chaotic, even a slight change in temperature or humidity can result in completely different weather. While day-to-day weather can feel unpredictable, patterns do form over time, helping us understand seasonal behaviors and make reasonable forecasts.

Similarly, the stock market often seems like a chaotic landscape, filled with random ups and downs caused by countless factors—from investor emotions to global events. Yet, if we take a closer look, we can spot underlying patterns that guide investment strategies. Technical analysis, for example, looks at historical price data and trading volumes to identify trends and make predictions. While stock movements can feel random in the short term, they often reveal a more structured reality over longer periods, shaped by economic forces.

The concept of randomness leading to predictability gets clearer when we explore mathematical ideas like the Law of Large Numbers and the Central Limit Theorem. The Law of Large Numbers tells us that as we conduct more trials, the average of the results will get closer to the expected value. So, if you were to flip a coin many times, you'd find that about half of the flips will be heads and half will be tails. What seems random at first becomes more predictable when we look at a larger sample size.

The Central Limit Theorem builds on this by stating that regardless of the distribution of the underlying data, the sample means will tend to form a normal distribution if the sample size is big enough. This theorem is crucial in statistics and explains why we often see the bell curve. The randomness of individual events can lead to predictable outcomes when we apply the Central Limit Theorem.

To make these concepts easier to understand, let's look at genetics and evolutionary biology. The variety we see in species comes largely from genetic variation, which is fundamentally random. Things like mutations, genetic drift, and recombination introduce variations that may seem chaotic. Nevertheless, these random genetic changes are critical to evolution, enabling species to adapt to their surroundings over time. Natural selection zeroes in on these random variations, favoring those that offer advantages and leading to the development of complex traits and behaviors.

In this light, randomness isn't just a hassle or a source of confusion; it's a key player in innovation and adaptability. A vivid example is the peppered moth in England, which changed its coloration during the Industrial Revolution. At first, lighter-colored moths thrived, blending in with lichen-covered trees. But as pollution darkened the environment, a random mutation resulted in the appearance of darker moths. This change gave them better camouflage from predators, leading to a higher survival rate for the darker moths. Here, we see how randomness and natural selection work together to create a clear and predictable outcome—a shift in the population.

Everyday stories of randomness also help illustrate these ideas. Think about the stock market, where you might invest in a company only to see its stock price shoot up or drop drastically overnight. While daily changes can feel random and chaotic, they often reflect larger trends influenced by economic data, consumer behavior, and global events. Over the long run, smart investors usually see their portfolios grow, benefitting from the fundamental principles of supply and demand—even if the path is bumpy.

Randomness also plays an important role in creative endeavors. The creative process often includes a level of unpredictability, whether through sudden bursts of inspiration or happy accidents. Artists, writers, and musicians frequently find that some of their favorite works arise from moments of serendipity, where chaos inspires creativity. The beauty of art captures the essence of randomness, turning fleeting moments into lasting experiences.

In science, randomness is embraced rather than rejected. Researchers design experiments to account for random variables, recognizing that uncertainty is part of the equation. The data gathered from these experiments leads to discoveries that enhance our understanding of the universe. So, randomness isn't something to fear; it's an important ally in the pursuit of knowledge.

As we think about the complex nature of randomness, we see that it's not just an abstract idea confined to math and science. It's a fundamental part of our lives, shaping our experiences, influencing our choices, and affecting the world around us. By welcoming the mysterious dance of randomness, we can learn to find meaning and order in life's unpredictability.

Randomness is a rich and layered phenomenon that goes beyond mere chance. It intertwines with order, crafting the patterns we see in nature, society, and our personal lives. From the development of a sunflower to the ebb and flow of the stock market, randomness acts as a creative force, spurring innovation and transformation. By grasping the essence of randomness and what it means for us, we can navigate through a world that often feels chaotic and unpredictable, finding comfort in knowing that even the wildest storms can lead to moments of clarity and beauty.

Philosophical Musings

As we think about the complex nature of existence, we're often led to consider the choices we make and the forces that shape our lives. The debate of free will versus determinism has been a long-standing topic in philosophy. It draws us in to examine whether our actions come from conscious decisions or if they are simply the results of a series of predetermined events. This question is fascinating because it raises a big dilemma: if everything we do is somehow destined, can we really say we have control over our lives?

Picture the brilliant mind of Pierre-Simon Laplace, an important figure from the Enlightenment era. He proposed that if we could know every particle in the universe and the laws of motion governing them, we could predict the future with complete accuracy. Every thought, every heartbeat, would be clear and exposed—a grand cosmic equation that shows the fate of everything. This deterministic viewpoint paints a picture of a perfectly ordered universe, where free will seems just a trick of our limited perspective.

However, as we dig deeper into the mysteries of existence, modern ideas from quantum mechanics challenge this deterministic view. At the subatomic level, particles behave in ways that are full of uncertainty. The well-known principle of superposition tells us that particles can exist in several states at once until they are observed. This introduces a layer of randomness that contradicts Laplace's confident predictions.

This quantum perspective makes us reconsider: if our choices come from a mix of fate and chance, what does that say about our ability to think and reflect?

This philosophical puzzle naturally extends into the world of creativity. We often romanticize the idea of the "eureka" moment. An artist staring at a blank canvas, a writer struggling with an empty page, or a composer facing silence are all engaged in a delicate balance of intention and spontaneity. The Infinite Monkey Theorem captures this idea perfectly. Imagine a monkey at a keyboard, randomly striking keys, and yet, somewhere in the vastness of infinity, it could accidentally type out the complete works of Shakespeare. What a delightful thought! This absurd scenario encourages us to think about how creativity often springs from chaos, good fortune, and unexpected combinations of ideas.

When we think about creativity, it's intriguing to realize how often artists tap into randomness to create their masterpieces. There's something beautiful about embracing the unpredictable nature of creativity—like an artist splashing paint without a plan, letting the colors mix and clash in surprising ways. While we admire the brilliance of careful design, there's also value in the happy accidents that emerge when we let go of control. Perhaps in these chaotic moments, true innovation comes to life, revealing secrets whispered by the muses between thought and chance.

Thought experiments abound in this arena, inviting us to break down the very essence of creativity. How much of our artistic expression results from careful planning? And how much comes from the whimsical randomness of our experiences? Think about a collage artist who collects discarded materials—old magazines, fabric scraps, photographs—and in what seems like a random arrangement, creates a story that resonates with viewers. Is this process simply a stroke of luck, or is it a thoughtful search for meaning? The reality is probably somewhere in the middle.

When we consider the nature of existence, it becomes clear that our understanding of infinity and randomness deeply influences how we perceive life and knowledge. Imagine standing at the edge of a vast ocean, with waves rolling in and out unpredictably. The endless horizon stretches before you, reminding you of the universe's bewildering complexity. Each moment, each choice, is like a single droplet in the ocean of existence—small but significant, swirling in currents that encompass everything.

This perspective encourages us to think about how randomness weaves through our experiences. Every encounter, every decision carries potential outcomes that come not only from our intentions but also from the unpredictable fabric of life itself. The experiences that shape who we are often result from a complicated mix of chance and choice. Our relationship with randomness can help us navigate life's uncertainties, showing us that even the most chaotic moments can lead to deep understanding.

With this insight, we can't help but ask questions. How do we balance the randomness of life with our desire for control? Are we just passengers on a cosmic journey, or do we have the power to steer our own ship through these currents? As we explore the intricate pathways of thought, we find ourselves not only reflecting on creativity but also on the essence of our very existence.

The investigation of randomness and creativity doesn't provide simple answers; instead, it sparks a meaningful conversation with significant implications. If creativity is an interplay of intention and chance, it raises questions about what it means to be an artistic genius. Are great artists just those who have learned to embrace the unpredictable? Or are they more like skilled surfers, riding the waves of chaos and expertly navigating the changing tides of inspiration?

We can also view this philosophical inquiry through the lens of infinity. Infinity isn't just a mathematical idea; it's woven into our understanding of creativity and existence. The countless combinations of ideas reflect the infinite possibilities of life. Just as a single brushstroke can resonate through time and inspire future generations of artists, a fleeting thought can shift the course of many lives.

As we consider these themes, we can't help but wonder how we could apply our understanding of randomness and creativity to our own lives. How can we nurture a mindset that welcomes uncertainty and allows for the unexpected? Perhaps the answer lies in letting go of the need for strict control and learning to dance with the uncertainties around us. Just like a jazz musician improvising and responding to the rhythms of their fellow musicians, we too can learn to adapt to the symphony of our experiences, allowing creativity to flow from both intention and spontaneity.

In the end, as we familiarize ourselves with the ideas of infinity, randomness, and creativity, we must see that these philosophical inquiries are more than just intellectual exercises. They are pathways to deeper insights about ourselves and our place in the universe. By embracing the unpredictability of existence, we open ourselves to a richer understanding of life— one that celebrates the beautiful chaos of creativity and the endless possibilities that arise in the dance between randomness and intention.

As we think about our beliefs and assumptions regarding creativity, free will, and the complex nature of reality, we find ourselves engaged in an ongoing conversation with life itself. In this space of reflection, we might ask ourselves: How much of our creative work comes from hard work, and how much comes from the delightful dance of chance? How can we use the power of randomness to enrich our experiences, allowing spontaneity to enhance what we create?

These questions invite us to reflect on the delicate balance between randomness and creativity, nudging us to embrace the uncertainties of life. Life, with all its unpredictability and infinite possibilities, offers us a canvas of potential waiting to be explored. As we delve into this philosophical exploration, we may find ourselves drawn to the beauty of the unknown, discovering that within the chaos lies the potential for profound understanding and revelation.

Chapter 9: The Dartboard Paradox – Zero Chances but Certain Outcomes

Understanding Zero Probability

When we start to think about probability, it can feel like we're lost in a confusing maze of numbers and odds. One idea that often trips us up is zero probability. It's puzzling because it makes us question what we believe about outcomes and events. How can something have a probability of zero yet still happen? It seems contradictory, right? How can an event that seems impossible actually occur? To understand this mystery, we need to wade through the tricky waters of continuous distributions and the ways our instincts can sometimes lead us astray.

Picture a dartboard, the kind you might find in a cozy bar or a game room. This dartboard has countless spots, each one waiting for a dart to land on it. If we try to figure out the chance of a dart hitting a specific point on this board, it might seem like the odds are totally against it. After all, a dartboard isn't just a flat surface; it stretches out infinitely with endless points. From a statistical viewpoint, the chance of a dart hitting any single, tiny point on the board is marked as zero. Yet, when you actually throw that dart, there's still a tiny chance—however unlikely—that it might land right where you aimed.

This puzzling situation comes from the nature of continuous distributions. When we discuss these in statistics, we're diving into a realm where outcomes aren't just separate, countable events. Instead, there's a whole range of possibilities. While it might seem logical to think that a probability of zero means something can't happen, the truth is more complex. The infinite nature of our dartboard shows us that, while the chance of hitting one exact spot is zero, the wealth of possibilities around that spot creates a fascinating mix of outcomes.

To clear this up, let's think more about density. Each point on the dartboard, although it has a tiny chance of being hit, exists within a larger set of possible outcomes. We can think of this density as how closely the points are packed on the board. Even if the chance of hitting one particular area is technically zero, the fact that there are countless points nearby, each with equal potential, means we can still land close to those areas. This sets up a paradox—while we can't hit that exact point, nothing stops us from landing very near it.

As we explore zero probability, we need to be careful about how misleading our instincts can be. Many people equate zero probability with something being impossible, but that's a big misunderstanding. To highlight this, let's consider a thought experiment involving a needle dropped onto a floor filled with infinitely many identifiable points. The odds of that needle landing precisely on any one point are indeed zero. However, the needle can—and will—land somewhere on the floor. The heart of this paradox is recognizing that while the needle's chance of touching a specific point is nil, it is completely possible for it to touch the floor anywhere else.

This concept of zero probability pops up in interesting stories and real-life examples all around us. Take the lottery, for example. When we buy a ticket, it feels like we're entering a grand game of luck. Statistically, our chances of winning are incredibly slim. Yet, every week, someone's life changes dramatically when they get that life-changing phone call announcing they've won. Just because the odds are so low doesn't mean winning is impossible; it shows how events with a seemingly zero probability can actually happen, shaking up lives in unexpected ways.

As we think about these examples, it's clear that grasping zero probability requires us to challenge our old beliefs and shift our views. The dartboard, the needle, and the lottery all remind us that the world is often much more complicated than the simple numbers we may initially consider. They encourage us to approach probability with an open mind, ready to embrace the paradoxes that lie within. Just because something seems impossible doesn't mean it can't occur.

In the vastness of our universe, where infinity stretches endlessly, outcomes aren't always as neatly defined as we might hope. The Dartboard Paradox invites us to think about how our beliefs shape our understanding of probability and, ultimately, reality itself. By embracing the complexities of zero probability, we open ourselves up to a richer appreciation of the infinite possibilities that come with each moment. This perspective helps us navigate the uncertainties life throws our way, reminding us that sometimes, the most unlikely events are the ones that profoundly shape our journeys. The insights hidden within the puzzle of zero probability not only intrigue us mathematically but also serve as a powerful metaphor for the unpredictable nature of our existence.

Infinite Possibilities

As we dive deeper into the world of probability, we arrive at a fascinating idea that's hard to pin down: the concept of infinite possibilities. Here, we leave behind the limited ideas of zero chances and step into a universe buzzing with outcomes that stretch beyond what we can imagine. Probability isn't just a simple "likely" or "unlikely" game. It's a rich and complex mix, filled with countless threads, each representing a potential event waiting to happen.

To make sense of this intricate idea, we first need to understand the role of density in probability, especially when we're looking at continuous spaces. Picture that dartboard once more—not just a flat piece of wood with paint, but a doorway into understanding probability itself. When we throw a dart, we're not just aiming for a single spot; we're interacting with a vast landscape of possibilities—a surface dotted with infinite tiny points. Each spot on the dartboard has a minuscule chance of being hit, but when we look at the whole board, we can assign a measurable, non-zero chance to bigger outcomes.

In probability, density is a key concept. It describes how outcomes are spread across a space. Instead of zeroing in on just one spot on the dartboard, we can think about larger sections of it. A certain part of the board can be hit, even if any specific point within that section seems unreachable. This idea is crucial because it reveals a paradox: even if individual outcomes have a zero probability, the density of nearby possibilities creates a lively universe of outcomes.

Let's imagine throwing a dart at the dartboard again. Visualize the dart soaring toward the board, aiming for that elusive bullseye. Each time it approaches, it has a zero chance of landing on any exact dot of paint. Yet, it has a good chance of hitting the surrounding area. The space around the bullseye is quite large, making it clear that the continuous nature of the dartboard allows for many outcomes, all nestled within the same infinite space.

Now, let's turn our attention to the infinite points on the dartboard. Each point, while theoretically singular, plays a role in the larger picture of probabilities. Think about the endless moments in time, with each moment serving as a potential point for events to happen—a day, an hour, a minute, a second. Time flows continuously, filled with countless moments, and it's easy to see how these individual moments can lead to very different outcomes. For example, if a runner decides to take a step at a certain moment, that single action could lead to winning a race or stumbling over a pebble. Each tiny moment is important, carrying implications and consequences that ripple through the world around us.

The idea of infinity can feel overwhelming, but when we connect it to our everyday lives, it becomes easier to grasp. Think about the countless books that come out every year. Each one is unique, yet together, they form a nearly infinite library of stories, ideas, and experiences. If we consider the chance of any one person reading a specific book among so many options, the odds seem tiny. Nonetheless, each reader's choice shows how infinite possibilities can emerge from seemingly unlikely events. This web of choices creates an exciting landscape where new worlds are always being explored, and fresh stories are born.

When we look at the infinite possibilities in our universe, we start to notice patterns in nature that reflect this unpredictable dance of events. Take mutation in biological evolution, for example. While the probability of any specific mutation happening in a population might be extremely low, the vast number of organisms and their interactions with the environment create nearly infinite chances for change. A single mutation, once considered a rare event, can trigger significant shifts in an ecosystem, leading to considerable evolutionary changes over long periods. The delicate balance of chance and survival shows how improbable events can lead to extraordinary outcomes, sometimes revolutionizing entire species.

This concept also connects to the world of quantum mechanics, where the behavior of particles embodies the essence of infinite possibilities. At the quantum level, particles don't just sit at one spot nor follow a predictable path. Instead, they exist in a cloud of probabilities, swirling with potential outcomes that challenge our traditional understanding of certainty. A particle can be here, there, or anywhere in between, with chances that go beyond simple yes or no answers. It's a realm where probabilities mingle and interact, creating an astonishing array of realities.

As we ponder these ideas, it becomes evident that infinite possibilities enrich our understanding of life itself. The Dartboard Paradox reminds us that our views can often limit our appreciation for what might be possible. Just because we can't pinpoint the chances of a specific event occurring doesn't mean the vast landscape of potential outcomes isn't swirling around us. This insight encourages us to embrace uncertainty and explore the endless opportunities it brings.

The infinite possibilities we encounter aren't just abstract ideas tucked away in the world of math and physics. They shape our daily lives, influencing our experiences and decisions in meaningful ways. From the simple choice of what to have for lunch to the complex decisions we face throughout our lives, each choice opens the door to a range of possible futures. Life unfolds like a branching tree, where every decision leads to new paths filled with their own unique possibilities, some of which might surprise us while others resonate deeply with who we are.

By being open to the infinite possibilities ahead of us, we nurture a sense of curiosity and wonder, allowing us to approach life with fresh eyes. Rather than fearing the unknown, we can see it as a blank canvas where we can create our own stories and discover the many paths that await us. The universe, with all its complexity, gives us the opportunity to witness the remarkable outcomes that can arise from what once seemed unlikely.

Ultimately, infinite possibilities remind us that life is a journey filled with uncertainty, yet brimming with chances for growth and discovery. They invite us to step beyond our comfort zones, challenge our assumptions, and open ourselves to the unexpected. In this grand adventure, we find meaning, joy, and connection. So, as we throw our darts into the vastness of life, let's do so knowing that every throw carries the potential for something amazing—a reminder that while the odds may look unfavorable, the infinite possibilities of life are waiting for us, just beyond the edge of the dartboard.

Applications in Science

As we move from the world of endless possibilities and rare chances, it's vital to think about how these ideas weave into the reality we experience every day, especially in science. At the heart of this exploration is quantum mechanics, a fascinating field that uncovers the unpredictable nature of the universe at its most basic level. The Dartboard Paradox, with its striking image of outcomes and probabilities, acts as a powerful metaphor for understanding the complexities of quantum events. When we picture the dartboard, we're not just imagining a game but contemplating the essence of existence itself—where every throw, every decision, symbolizes the unpredictable, often surprising behavior of the tiny particles that make up our world.

In the quantum realm, uncertainty is the name of the game. This uncertainty isn't just a theoretical idea; it's a fundamental principle in the laws of physics. Here, events that might seem impossible actually happen more often than we think. Electrons don't sit still like marbles on a table; instead, they exist in a cloud of probabilities, with their positions and speeds described by mathematical equations rather than certainties. The dartboard serves as a lively model for this, where every point represents a potential spot for an electron, yet our ability to pinpoint its exact location at any moment remains tricky.

The core of quantum mechanics rests on the ideas of probability and uncertainty. Heisenberg's Uncertainty Principle captures this beautifully. It tells us that the more accurately we know an electron's position, the less accurately we can know its momentum, and the other way around. This limit is a key part of the quantum world. Imagine throwing a dart at the board while also trying to measure how fast it's going and where it lands. The more focus you put on one aspect, the more unclear the other becomes. It's a delicate dance of uncertainty that not only defines how subatomic particles behave but also shapes our understanding of reality itself.

This idea flips our traditional views upside down. In the larger world, we're used to a clear cause-and-effect: push a ball, and it rolls in a straight line. In contrast, the quantum world operates with probability waves, where what we see is just one of many possible results. A great example of this is how photons behave when passing through a double-slit apparatus. When light hits two slits, it can create an interference pattern, as if each photon is acting like a wave, taking both paths at the same time. However, if we try to figure out which slit the photon goes through, it mysteriously acts like a particle, collapsing the probability wave into one clear outcome. Just like our dart, its journey changes based on whether we're watching, which shows the complex relationship between the observer and what's being observed.

Think about what this means. At its core, the idea that things can exist in multiple states until we measure them connects closely to the Dartboard Paradox. Each possible spot of the dart on the board is similar to the possible states of a particle before we observe it. This realization leads us down a path of questions about reality itself: What does it mean for something to "exist" if no one is watching? Could it be that our grasp of existence is limited by our own perspectives? The dartboard, symbolizing our understanding, is filled with potential outcomes waiting to become real, encouraging us to rethink how we view the universe around us.

As we dig deeper into the world of quantum mechanics, we discover real-world applications that bring the principles of the Dartboard Paradox to life. Take quantum computing, for instance. The power of a quantum computer lies in the concept of superposition, where a qubit (quantum bit) can exist in multiple states at the same time. Imagine a dart that can hit several spots on the dartboard all at once—this captures the essence of quantum computing. It allows for operations on many inputs simultaneously, making it vastly more efficient than traditional computers for certain tasks. The potential of quantum computers is in their ability to explore possibilities that classical computers simply can't, leading to breakthroughs in fields like cryptography, drug discovery, and complex system modeling.

In the fascinating world of quantum entanglement, we find another vivid example of the Dartboard Paradox. Entangled particles are linked in such a way that the state of one instantly impacts the state of another, no matter how far apart they are. This phenomenon challenges our traditional views of space and cause-and-effect, reminding us that the universe functions on principles that often go beyond our understanding. Visualize entangled particles as two darts thrown at unseen boards—the result of one dart directly affects the outcome of the other, even if they're light-years away. This connection suggests a deep unity at the core of our universe, echoing the threads of infinite possibility.

Heisenberg's Uncertainty Principle, quantum computing, and entanglement together showcase how the Dartboard Paradox moves beyond theoretical ideas to influence our scientific understanding. Each of these concepts highlights a part of reality where unpredictability rules, and a multitude of possibilities lies just below the surface. They push us to rethink our perceptions, inviting us to embrace uncertainty and discover the wonders of the quantum world.

Yet, the Dartboard Paradox doesn't navigate this journey alone. It resonates with the philosophical questions that have been part of scientific inquiry throughout history. The search for understanding often leads us to places where certainty slips away, and probability takes center stage. As we explore the intricate landscape of quantum mechanics, we start to see that the uncertainties we face aren't obstacles but gateways to a deeper understanding.

You might wonder how this significant uncertainty plays out in our everyday lives. The answer lies in the ways we engage with the universe daily. From the ordinary moments to the big decisions, our lives are influenced by probabilities, many of which we don't consciously notice. Think about making a choice—whether to cross the street or take a new job. Each choice carries countless potential outcomes, like a dart thrown toward a dartboard of infinite possibilities. The hidden forces of probability guide us, leading us down paths that might surprise us and reminding us that every step we take interconnects with the vast web of choices around us.

The Dartboard Paradox encourages us to welcome uncertainty, not just as a scientific concept but as a guiding principle in our lives. As we grapple with the complexities of existence, we are inspired to stay open to the many possibilities that lie ahead. Life itself is a journey filled with choices, interactions, and the unpredictable nature of chance. By recognizing the probabilistic nature of our reality, we are better prepared to navigate the uncertainties that life throws our way.

As we think about the deep connections between zero probability, infinite possibilities, and their influence on scientific exploration, it becomes clear that the Dartboard Paradox is more than just a mathematical curiosity. It invites us to rethink what we know about reality, encouraging us to embrace uncertainty as a pathway to growth and discovery. The universe, in all its complexity, is a rich landscape of outcomes just waiting to be explored. Within the realm of the improbable, anything can happen.

Ultimately, the Dartboard Paradox reflects the foundations of quantum mechanics and the essence of our existence.

Avery Pascal

Our understanding of probability and uncertainty, rooted in both mathematical principles and philosophical questions, gives us a glimpse into the intricate dance of chance that shapes our lives. It challenges us to acknowledge the unknown, to explore the infinite complexities that await us, and to appreciate the boundless possibilities that lie just beyond the edge of the dartboard. In this extraordinary universe, where probability reigns supreme, the only certainty is that uncertainty will always be by our side.

Chapter 10: The St. Petersburg Paradox – Betting on Infinity

The Game Setup: Understanding the St. Petersburg Game

Imagine sitting in a cozy room with a few friends, maybe sipping a drink, and feeling the excitement of friendly competition. One of your buddies, a math whiz who loves quirky ideas, brings up a game that sounds almost too wild to believe. He says it offers a chance to win huge amounts of money, all based on a simple coin flip. This game is called the St. Petersburg Game, and it mixes probability, risk, and the thrilling idea of limitless wealth.

The game starts off in a way that pulls you in. To join in, you have to pay a fixed entry fee—let's say $10. With that small investment, you're off on your adventure towards possible riches. The rules are easy to grasp: a coin gets flipped. If it lands on heads, the game continues, and you earn a payout that doubles with each heads flipped. The stakes rise quickly. One flip gives you $2, two flips give you $4, three flips give you $8, and so on. If you keep getting heads, your winnings could skyrocket, leading you towards the idea of infinite wealth.

Now, let's break down the math behind the game. The expected value of this game is figured out by looking at all possible outcomes from the coin flips. For every (n) heads flipped in a row, the payout is (2^n). You can calculate the expected value (EV) using the formula:

$$ EV = \sum_{n=1}^{\infty} \left(\frac{1}{2^{n+1}} \cdot 2^n \right) $$

Here's how it shakes out: each time you flip the coin, you have a 50% chance of winning that round's payout. If you flip heads once, you win $2; twice, you get $4; three times, $8; and so on. The chance of flipping heads (n) times in a row is ($\frac{1}{2^{n+1}}$), which means you need to get heads (n) times and then tails to stop. When you add up this infinite series, it leads to an expected value that seems infinite.

This brings up a puzzling question: why don't more people leap at the chance to play a game where the expected payout is infinite? The answer lies in the gap between math and how people feel. Even though the numbers suggest endless possibilities, the reality of risk and reward makes many players cautious. The thought of losing that initial $10 can feel daunting, overshadowing the tempting idea of winning big.

To illustrate this feeling, let's picture a sports betting situation. You might find yourself at a bar, cheering for your favorite underdog team against the reigning champs. You decide to place a $20 bet on your team to win. If they manage to pull off the upset, you could double your money or more. What draws you to this bet isn't just the odds; it's the emotional connection—the thrill of rooting for your team and the excitement of competition. Compare this to the St. Petersburg Game, where everything hinges on the unpredictability of a coin flip. That emotional disconnect can be really tough to deal with.

    Now think about investing in the stock market. You might put money into a promising startup, excited about the chance for big growth. If the company succeeds, your small investment could multiply impressively. But the fear of losing your money is very real. Just like with the St. Petersburg Game, the stakes feel much more serious in the real world, where real investments and emotional ties are involved.

    For many people, their reluctance to join in on the St. Petersburg Game can be linked to the idea of diminishing returns. Even if the theoretical payout is limitless, the practical side of things feels much more grounded. Players often conclude that they'd need to flip heads an improbable number of times to actually get those huge payouts.

These mental barriers keep them from participating, even while the numbers seem to tell a different story.

Real-life stories from gamblers add more depth to this concept. Imagine a seasoned poker player, used to the thrill of high-stakes games, sitting across from a newbie at a casino table. The novice might be excited by the chance to win big, but they can quickly become overwhelmed by the game's complexity and unpredictability. On the flip side, the poker player has learned how to manage risks, read their opponents, and understand probabilities in ways that the novice still doesn't grasp. For the poker player, the excitement comes not just from the potential rewards but from mastering the strategies of the game.

The St. Petersburg Game shines a light on the difference between what the math suggests and how people actually behave. Even though the expected value calls out for attention, players remain careful, struggling to fit the idea of infinity into their own cautious mindset. This puzzle opens the door to deeper discussions about economic behavior, decision-making, and how we think about risk.

As we explore this fascinating game, we need to face the tricky questions that come from our natural discomfort with risk. How do we balance what's theoretical with what's real? What does it mean to take a gamble in a world where the stakes can seem too high? The St. Petersburg Game, with its promise of infinity, pushes us to rethink how we relate to uncertainty and reward, inviting us to look closely at our own decision-making habits, rather than just getting swept away by the idea of unlimited winnings.

In this intricate dance between math and psychology, the St. Petersburg Game serves as a powerful reminder of the complexities of human behavior. It encourages us to dig into not just the mechanics of the game, but also the reasons behind our choices. As we unravel this puzzle, we'll find that the lessons we learn go beyond gambling, reaching into the very core of how we handle risk in our daily lives.

Expected Value vs. Reality: The Clash Between Theory and Practice

The St. Petersburg Game is a fascinating idea, a tempting lure of infinite expected value that promises endless wealth. However, when it's time to actually play, people often hesitate and choose safer options that seem to contradict the math behind the game. This puzzling situation raises important questions about how we understand risk, decision-making, and what drives human judgment.

At the core of this conversation is expected utility theory, which helps explain why rational people often decide against playing the St. Petersburg Game, even though the math looks appealing. Sure, the expected value of the game is infinite, but the reality of that potential is clouded by our natural fear of risk and the fact that the satisfaction we get from money decreases as we have more of it.

Let's break this down with a simple truth about how we behave: we don't see money in a straight line. The first dollar you earn feels more important than the hundredth. This idea is called diminishing marginal utility. It means that as you have more money, the extra joy (or utility) you get from each additional dollar isn't as great. So, while the St. Petersburg Game promises a huge payout, the joy you'd get from winning that money might not be enough to make you ignore the upfront cost and the chance of losing it.

Picture this: you're deciding whether to play the St. Petersburg Game. You have $10, which is the entry fee, but you also need to pay your bills and buy groceries. The potential winnings are exciting, but so is the worry of losing that $10. If you win, your payout could be huge, but think about the mental toll of losing that initial investment. For many, the fear of losing that money overshadows the excitement of possibly winning. This is where risk aversion comes into play. People generally prefer to avoid losses rather than focus on gaining something. It just hurts more to lose something you already have than to think about the thrill of winning something new.

Now, let's look at lottery tickets—another situation where expected value and our behavior don't quite match up. Lottery players are often attracted to the chance of life-changing jackpots. They spend small amounts on tickets, fully aware that their odds of winning are slim. The expected value of most lottery games is pretty low, often below the ticket price, yet people still rush to buy them. Why? The excitement of possibility, the hope of winning, and the fun of being part of something that grabs everyone's attention often matter more than a cold, hard analysis of the odds.

Think about your last visit to a casino. You might have walked in with a budget you planned carefully, ready to enjoy some games of chance. But once you're there, the energy in the room and the prospect of hitting it big can sway your choices. Your decisions may be driven more by your emotions, biases, and the influence of others rather than a strict focus on expected values. You might find yourself drawn to a slot machine or poker table, even when the odds are against you, caught up in the thrill and excitement of gambling.

This behavior can be partly explained by cognitive biases, like the availability heuristic. This is where people judge the likelihood of events based on how easily they can recall examples. If a friend won a big poker game last month, that memory can make you feel more optimistic about your chances, even if the odds remain the same. The high from that win can overshadow the reality of the risks, leading you to approach the game with hope that isn't backed by the actual statistics.

In contrast, the St. Petersburg Game is straightforward math but lacks the emotional connection found in other gambling scenarios. It's just a gamble based on a coin flip, stripped of the rich stories and personal ties that often come with other games. Even though the game promises infinite rewards, it feels distant and abstract. The randomness that should attract players can instead create a mental barrier. The starkness of the game forces you to confront the harsh reality of chance, which can be off-putting.

Moreover, the choice to play the St. Petersburg Game reaches beyond just money. It touches on self-perception and identity. Playing the game might lead you to question your financial skills or risk your self-image on something that seems irrational. There's also the societal pressure that comes with gambling—like the shame of losing—which can make people reluctant to even risk a small amount on a game with uncertain outcomes.

As we dive into these themes, we uncover layers of human psychology. The St. Petersburg Game serves as an intriguing example that shows just how irrational our decisions can sometimes be. It brings to light how our feelings, biases, and social situations can heavily influence our choices, often leading us to make decisions that seem contradictory or illogical from an expected value perspective.

In behavioral economics, understanding the gap between expected value and human behavior has real significance. Our choices are shaped not just by cold calculations but also by our feelings. This interplay invites us to rethink how we design economic systems, marketing campaigns, and policies that rely on decision-making.

Let's pause and think about what this means. If we see that people often act against traditional economic predictions, how can we use this knowledge? For example, marketing for high-stakes games or lotteries might resonate more with the emotional reasons people play instead of simply highlighting potential financial gains. Recognizing the importance of storytelling and the social side of risk could lead to better strategies that align more closely with how people really make choices.

The St. Petersburg Game and the paradox it presents aren't just academic concepts; they reflect our everyday experiences and decisions. They invite us to dig deeper and understand the complexities of human psychology. They urge us to rethink the reasons we take risks, how we handle our fears, and what it means to face uncertainty in a world that often feels unpredictable.

Ultimately, the dance between expected value and reality is complex and rich with meaning. It encourages us to explore the psychological factors that guide our choices, pushing us to face not only the allure of infinite returns but also the reasons we might hesitate to chase after them. The St. Petersburg Game gives us a glimpse into infinity, but it's in the messy, unpredictable world of human emotions that we truly discover the essence of decision-making. As we navigate this journey, we begin to appreciate the intricate nature of being human, where rational thought and feelings coexist, often leading us down paths that defy the straightforward math. In this exploration, we uncover the intricacies of our choices, ultimately realizing that the paradox of infinity is not just an intriguing math problem but a deep reflection on the human experience itself.

Economic and Psychological Insights: Implications of the Paradox

When we think about the St. Petersburg Paradox, we often find ourselves grappling with the idea of endless possibilities set against our deep-rooted desire for safety and security. At first glance, this puzzle seems like just a simple math problem, a tricky riddle that teases our brains. But when we dig a little deeper, it unfolds into a fascinating look at our economic behaviors and the psychological factors that drive them. This exploration helps us understand how we navigate the twists and turns of risk, rewards, and the personal interpretations of value that shape our everyday choices.

To truly grasp what the St. Petersburg Paradox means, we need to break down the details of expected utility theory. While the expected value gives us a clear mathematical outcome—a limitless payout—the truth about human behavior is much more intricate. When people make choices, they don't just crunch numbers; they also consider their emotions, experiences, and the context of the situation. The gap between expected value and expected utility captures our struggle with the St. Petersburg Game and reflects the wider challenges we encounter in making economic decisions.

Expected utility theory suggests that individuals look at potential outcomes not just through the lens of expected value, but also based on their individual experiences and personal preferences. Each person brings their own set of emotional weights to the table—factors such as fear of losing, how choices are framed, and the impact of past experiences. So, while one person might focus solely on the infinite expected return of the St. Petersburg Game, another might be more concerned about the emotional hit of potentially losing that initial entry fee. The choices we make can seem illogical, especially when viewed through the traditional economic lens, which often assumes that everyone is perfectly rational and only acts in their best interest.

Imagine for a moment choosing between playing the St. Petersburg Game or taking a guaranteed $10. The logical argument would lean toward the game, given its infinite expected value. However, the psychological reality tells a different story. Many people would go for the guaranteed $10, a choice rooted in the comfort of certainty and the desire to avoid loss. This behavior, known as loss aversion, highlights how the pain of losing something feels more intense than the joy of gaining something equal in value. The paradox, then, reveals not just a clash of numbers, but also a reflection of our emotional experiences.

One of the most striking takeaways from the St. Petersburg Paradox is how it showcases the limitations of traditional economic models. Classic economic theories often assume that people act in their own self-interests and make rational choices based only on numbers. Yet, the St. Petersburg Game serves as a reminder that we are not just mathematical entities; we are complex beings shaped by emotions, biases, and social influences. This complexity calls for a fresh approach to how we model economic behavior.

Now, think about how this affects policymakers and marketers. If our choices are influenced more by feelings than calculations, then creating economic strategies needs to take into account the emotional side of risk. Instead of just trying to persuade someone to invest based on potential returns, a more effective strategy might involve crafting a story that connects with their hopes and fears. Understanding the psychological triggers behind decisions can lead to better communication and marketing strategies, especially in high-stakes situations like investments, insurance, or charitable donations.

A vivid example of these ideas can be seen in the world of gambling. Casinos pour resources into building environments that ramp up excitement and emotional engagement. The bright lights, catchy sounds, and social atmospheres are all carefully designed to heighten the thrill of the game, often leading players to make decisions that contradict traditional expected value calculations. In the context of the St. Petersburg Paradox, this means that the lure of infinite returns can be overshadowed by the exhilarating experience of the casino. Players frequently chase the high of potential rewards without fully considering the odds, showing how emotional ties can lead to decisions that stray from rational expectations.

This pattern extends beyond gambling and seeps into many areas of our lives. We often witness individuals making spending choices that don't align with the expected value of products. For instance, someone might buy a luxury item with a relatively low expected return simply because the emotional connection to owning it outweighs the logical cost-benefit analysis. It's a clear illustration of how the St. Petersburg Paradox plays out in everyday life—where countless possibilities clash with the limits of our desires and fears.

In understanding the St. Petersburg Paradox, we also uncover the significance of personal preferences and how we perceive risk. What may seem like a trivial gamble to one person could represent a major dilemma for someone else. This difference is often seen in how individuals approach investments. Two people might encounter the same financial opportunity, yet their decisions could be worlds apart due to their unique experiences, risk tolerance, and emotional reactions to potential losses. The paradox encourages us to consider the various perspectives we adopt regarding value and risk, reminding us to appreciate the rich diversity of human experience as we make our choices.

As we explore further into the psychological implications of the St. Petersburg Paradox, we come across the idea of framing. The way options are presented can greatly influence our decisions, especially in uncertain situations. For example, if a lottery ticket is framed as a chance to win a life-changing amount of money, it might appeal to people more than if it were described as a risky gamble with slim chances of success. The framing effect shows just how easily our perceptions can be swayed, leading us to make choices based more on context than on hard facts.

On a larger scale, what we learn from the St. Petersburg Paradox pushes us to rethink our understanding of economic models. It encourages us to look at not just the math behind decision-making, but also the complicated web of psychological factors that affect our choices. The insights from this paradox go beyond economics and touch on the core of human behavior. They invite us to explore how we deal with uncertainty, how we establish value, and how we manage the unpredictability of life.

Reflecting on these ideas reveals an important truth: the St. Petersburg Paradox isn't just a mathematical oddity. It serves as a powerful lens for understanding the complexities of how we make decisions and our relationship with risk. It urges us to recognize that while numbers can guide us, it's our emotions, biases, and experiences that truly shape our choices.

With these insights in mind, it's clear that the implications of the St. Petersburg Paradox are wide-ranging. They prompt us to think about how we can apply these lessons in our own lives. As we navigate our personal and financial decisions, it's beneficial to stay aware of the psychological elements at play. By recognizing our biases, understanding our perceptions of risk, and being mindful of framing, we can empower ourselves to make wiser choices.

Furthermore, as we connect the paradox to our daily lives, we may find ourselves better prepared to tackle the uncertainties that inevitably come our way. The lessons learned from the St. Petersburg Paradox can serve as a roadmap, guiding us through the complexities of decision-making and helping us find significance in the unpredictability that life offers. As we confront our relationship with risk and reward, we not only clarify our own motivations but also foster a greater appreciation for the intricate dynamics that shape human behavior.

Avery Pascal

Conclusion

As we reach the end of our journey through the paradoxes of infinity, I hope you're feeling a mix of wonder, curiosity, and perhaps a healthy dose of mind-bending confusion. That's exactly how I feel every time I ponder these concepts, and it's what keeps me coming back for more.

Remember the infinite hotel we visited? Or the set that contained itself? These aren't just mathematical oddities – they're invitations to see the world differently. The next time you're faced with a seemingly impossible problem, think about Hilbert's Hotel. There might be a solution you haven't considered yet, hiding in the infinite possibilities.

Infinity isn't just a mathematical concept – it's a mindset. It's about pushing boundaries, questioning assumptions, and never stopping our quest for knowledge. So, what's next? Well, that's up to you. The journey doesn't end here – in fact, it's infinite!

Keep asking questions. Keep exploring. And most importantly, keep looking at the world with curiosity and wonder. After all, in a universe full of infinite possibilities, who knows what amazing discoveries await?

Avery Pascal

www.ingramcontent.com/pod-product-compliance
Lightning Source LLC
Chambersburg PA
CBHW052156220526
45471CB00004B/1698